AF581778

# Advanced Materials for Electrochemical Energy Conversion and Storage Devices

# Advanced Materials for Electrochemical Energy Conversion and Storage Devices

Editors

**Diogo M. F. Santos**
**Biljana Šljukić**

MDPI • Basel • Beijing • Wuhan • Barcelona • Belgrade • Manchester • Tokyo • Cluj • Tianjin

*Editors*

Diogo M. F. Santos
Center of Physics and Engineering of Advanced Materials
Laboratory for Physics of Materials and Emerging Technologies
Chemical Engineering Department
Instituto Superior Técnico
Universidade de Lisboa
Lisbon
Portugal

Biljana Šljukić
Center of Physics and Engineering of Advanced Materials (CeFEMA)
Instituto Superior Técnico
Universidade de Lisboa
Lisbon, Portugal and
Faculty of Physical Chemistry
University of Belgrade
Belgrade
Serbia

*Editorial Office*
MDPI
St. Alban-Anlage 66
4052 Basel, Switzerland

This is a reprint of articles from the Special Issue published online in the open access journal *Materials* (ISSN 1996-1944) (available at: www.mdpi.com/journal/materials/special_issues/adv_mater_energy).

For citation purposes, cite each article independently as indicated on the article page online and as indicated below:

LastName, A.A.; LastName, B.B.; LastName, C.C. Article Title. *Journal Name* **Year**, *Volume Number*, Page Range.

**ISBN 978-3-0365-3283-7 (Hbk)**
**ISBN 978-3-0365-3282-0 (PDF)**

# Contents

# About the Editors

**Diogo M. F. Santos**

Diogo M. F. Santos is an Invited Assistant Professor at Instituto Superior Técnico (IST, ULisboa) and Researcher in the Center of Physics and Engineering of Advanced Materials (CeFEMA), studying electrodes and membranes for application in direct liquid fuel cells. D.M.F. Santos got his PhD degree in electrochemistry in 2009, focusing on developing the direct borohydride fuel cell (DBFC). He did a postdoc in Chemical Engineering on hydrogen production by alkaline water electrolysis, followed by a postdoc in Materials Science & Engineering to develop cathodes for DBFCs. As an Investigador FCT in CeFEMA (2015-20), he worked on low-cost electrocatalysts for low-temperature fuel cells.

D.M.F. Santos has authored 120 journal papers and 90 conference proceedings, and his current h index is 30 (SCOPUS). He is in the "World's Top 2% Scientists list" of Stanford University for the impact in 2020. D.M.F. Santos has presented more than 50 oral communications and 80 posters at international conferences. He is a member of renowned international societies and serves as a reviewer for over 80 scientific journals. D.M.F. Santos is the supervisor of one PhD student and one MSc student. He has completed the supervision of one Postdoc researcher, one PhD student, 17 MSc students, and 10 research internships. His main research interests are related to electrochemical energy conversion and storage.

**Biljana Šljukić**

Biljana Šljukić Paunković graduated in physical chemistry at the University of Belgrade –Faculty of Physical Chemistry in 1999 and received M.Sc. degree from the same institution in 2003. She received a DPhil degree in chemistry from University of Oxford in 2007. She continued her academic career at the UB - Faculty of Physical Chemistry as an associate professor and as researcher at the Center of Physics and Engineering of Advanced Materials (CeFEMA), Instituto Superior Técnico, University of Lisbon.

Biljana's research is focused on novel materials for application in electrochemistry energy conversion and storage, and she has authored more than 100 papers, two book chapters, and one university book. Furthermore, she is/was coordinator of national and international projects in the area of electrochemistry energy conversion and storage on lithium-ion and metal-air batteries. Another field of her research is development of materials for electroanalytical applications and sensors.

Biljana has supervised six defended PhD, 25 Master, 29 Undergraduate, and 21 Vocational studies theses. Besides her research and pedagogical activities, Biljana is also very active in the area of science promotion, so she is/has coordinated projects within Horizon 2020/Horizon Europe - Marie Skłodowska-Curie Actions European Researchers' Night program.

# Preface to "Advanced Materials for Electrochemical Energy Conversion and Storage Devices"

Electrochemical energy conversion and storage is attracting special attention due to the drawbacks and limitations of existing fossil fuel-based technologies. Cleaner ways to produce and store energy are required to relieve our global dependence on highly polluting fossil fuels. This book focuses on recent developments in electrocatalytic materials for application in the broad field of energy storage and conversion. It brings the latest advances in the synthesis and characterisation of novel materials for electrochemical energy conversion and storage devices, including high-efficiency lithium-ion rechargeable batteries, supercapacitors, and alkaline water electrolysers. This book is expected to be a valuable tool for those from industry and academia interested in knowing more about the field. It includes contributions from academicians affiliated with research institutes from 14 different countries. The editors would like to express their sincere gratitude to those who made this book possible.

**Diogo M. F. Santos, Biljana Šljukić**
*Editors*

*Editorial*

# Advanced Materials for Electrochemical Energy Conversion and Storage Devices

**Diogo M. F. Santos [1,*] and Biljana Šljukić [1,2]**

[1] Center of Physics and Engineering of Advanced Materials (CeFEMA), Instituto Superior Técnico, Universidade de Lisboa, 1049-001 Lisbon, Portugal; biljana.paunkovic@tecnico.ulisboa.pt
[2] Faculty of Physical Chemistry, University of Belgrade, Studentski Trg 12-16, 11158 Belgrade, Serbia
* Correspondence: diogosantos@tecnico.ulisboa.pt

Electrochemical energy conversion and storage is attracting particular attention due to the drawbacks and limitations of existing fossil fuel-based technologies. Progress in electrochemical energy conversion/storage devices takes three directions: batteries, supercapacitors, and fuel cells. Batteries find wide applications in portable devices, including laptop computers, mobile phones and cameras. Supercapacitors can accept and deliver charge at a much faster rate than batteries, and for many charge/discharge cycles. Fuel cells provide efficient and clean continuous power generation for stationary and portable applications. These three technologies are considered to be clean and promise to overcome climate change problems caused by the intensive use of fossil fuels.

However, issues related to electrode efficiency, membrane costs, and electrolyte stability still often limit the widespread commercialisation of electrochemical energy conversion/storage devices. Namely, the choice of electrode materials, as well as the electrolyte composition, determines the crucial electrochemical device parameters, such as specific energy and power, cycle life and safety. Accordingly, it is essential to develop the existing and introduce new procedures for synthesising electrode materials for batteries, capacitors and fuel cells. Developing new, improved electrocatalytic materials for batteries, supercapacitors, and fuel cell electrode reactions is expected to significantly impact device performance and, consequently, their commercialisation.

The present special issue is focused on recent developments in electrocatalytic materials for energy storage and conversion devices. It brings the latest advances in the synthesis and characterisation of novel materials for electrochemical energy conversion and storage devices, including high-efficiency lithium-ion rechargeable batteries, supercapacitors, and alkaline water electrolysers.

Lithium-ion batteries are the primary energy storage devices in the communications and renewable-energy sectors due to their high energy densities and lightness. In addition, they have no memory effect and do not use poisonous metals, such as lead, mercury or cadmium. Still, further enhancements are necessary concerning battery power, cycle life, safety, and cost, to meet some applications requirements. The performance of batteries strongly depends on the three main parts: the electrodes (anode and cathode) and the electrolyte solution properties. Thus, significant improvement in the electrochemical properties of electrode materials is essential to meet the demanding requirements of various battery applications.

A lithium-ion battery commonly comprises a carbon anode, a metal oxide (a layered oxide, a polyanion, or a spinel) cathode and a lithium salt in an organic solvent as the electrolyte. A separator in the form of a thin sheet of micro-perforated plastic is commonly employed to separate the anodic and cathodic parts while allowing ions to pass through. This special issue presents several novel cathode materials for LIBs. Thus, Hussmes et al. prepared a series of $LiNi_{1/3}Mn_{1/3}Co_{1/3}O_2$ cathode materials (doped with Al, Mg, Fe, and Zn) using a rather facile sol-gel method assisted by ethylenediaminetetraacetic acid (EDTA) as a chelating agent [1]. X-ray diffraction (XRD) analysis revealed

**Citation:** Santos, D.M.F.; Šljukić, B. Advanced Materials for Electrochemical Energy Conversion and Storage Devices. *Materials* **2021**, *14*, 7711. https://doi.org/10.3390/ma14247711

Received: 6 December 2021
Accepted: 9 December 2021
Published: 14 December 2021

that the degree of cation mixing, as well as the strain field, depended on the doping element. Their morphology inspected by scanning and transmission electron microscopy (SEM/TEM) was regular with hexagonal nanoparticles of practically the same size allowing proper assessment and comparison of their electrochemical performance. Doping with Mg and Al resulted in enhanced electrochemical performance as evidenced by the enhanced specific capacity, cyclability, and rate capability, owing to the minimum occupancy of foreign ions in the Li plane. Specifically, doping with Al increased the initial capacity (i.e., delivered higher discharge capacity in the first cycles compared to the Mg doping), but doping with Mg improved the long-term cyclability. Initial specific discharge capacity at 0.1 C decreased in the order Al-doped oxide (160 $mAhg^{-1}$) > Mg-doped and undoped (ca. 150 $mAhg^{-1}$) > Fe-doped (146 mAh $g^{-1}$) > Zn-doped material (118 $mAhg^{-1}$); doping with Fe and Zn had a negative effect on the specific discharge capacity value. On the other hand, specific capacity retention decreased in the order Mg-doped oxide (91%) > undoped (85%) > Al-doped (82%) > Fe-doped (67%) > Zn doped material (36%). The improved capability was attributed to the minimised local distortions in the lattice.

The work of Ajpi et al. presents the synthesis of lithium iron phosphate–polyaniline ($LiFePO_4$–PANI) hybrid materials and their electrochemical performance as electrode materials for LIBs [2]. A high degree of crystallinity PANI was synthesised and optimised by chemical oxidation using ammonium persulfate and phosphoric acid. SEM analysis showed PANI formation with primary particles of 0.31 μm size and globular morphology with agglomerates of 2.75 μm. $LiFePO_4$–PANI hybrid was subsequently synthesised by thermal treatment of $LiFePO_4$ particles in a furnace with PANI and lithium acetate-coated particles under inert ($Ar/H_2$) atmosphere. The morphologies of $LiFePO_4$ particles and $LiFePO_4$–PANI hybrid were observed to be similar, whereas their structures were found to be different due to the presence of PANI coating on the $LiFePO_4$ particles. Elemental mapping confirmed a homogeneous distribution on the surface of the $LiFePO_4$ particles.

The presence of PANI in the hybrid material was shown to have a beneficial effect on the hybrids' electrochemical performance, i.e., on its rate capability. Namely, hybrid material with 25 wt.% PANI exhibited enhanced electrochemical behaviour in terms of capacity, rate capability and cyclability compared to the individual components, $LiFePO_4$ and PANI. Thus, capacity at a charge/discharge rate of 0.1 C increased in the order PANI (95 $mAhg^{-1}$) < $LiFePO_4$ (120 $mAhg^{-1}$) < $LiFePO_4$–PANI (145 $mAhg^{-1}$). The beneficial effect of PANI presence is also evident at a higher charge/discharge rate of 2 C, with a capacity of 70 and 100 $mAhg^{-1}$ for $LiFePO_4$ and $LiFePO_4$–PANI, respectively. PANI improved the electronic transfer between $LiFePO_4$ particles and the contact between $LiFePO_4$ particles and electrolyte during charge/discharge. It can further serve as $Li^+$ ion insertion–extraction host contributing to the capacity. Finally, PANI can act as a binder network between the $LiFePO_4$ particles and the surface of the current collector.

For fuel cells to be considered a clean technology, the hydrogen used in $H_2/O_2$ fuel cells must be produced by clean/green methods. Still, most of the $H_2$ produced globally represents the so-called "grey" hydrogen, i.e., it is obtained by natural gas reforming. Green hydrogen can be generated by water electrolysis powered with electricity from renewable energy sources such as wind or solar energy. The main problem of $H_2$ production by water electrolysis is its high price. This high cost comes from the high energy input, i.e., large overpotential needed to split the water molecule. The overpotential can be reduced by using suitable electrode materials, i.e., with high activity for the hydrogen (HER) and oxygen (OER) evolution reactions. Cysewska et al. explored a series of nanostructured Mn-Co-based films as electrode materials for oxygen evolution reaction [3]. These films were prepared using different synthesis conditions, which were correlated with the films' physicochemical properties (e.g., structure, morphology) and, consequently, with their electrocatalytic activity. Mn-Co nanofilm was directly electrochemically deposited in a one-step process on nickel foam from the solution containing only metal nitrates ($Mn(NO_3)_2{\cdot}4H_2O$ and $Co(NO_3)_2{\cdot}6H_2O$) and no additives. XRD and X-ray photoelectron spectroscopy (XPS) analyses revealed that the as-prepared films consist mainly of $Mn^{2+}$, $Co^{2+}$, and $Co^{3+}$ oxides/hydroxides and/or oxyhydroxides. The subsequent alkaline treatment of the film in 1 M KOH led to partial oxidation of $Co^{2+}$ to $Co^{3+}$ and generation of $Mn^{3+}$, resulting in Mn-Co oxyhydroxides. The creation of crystalline $Co(OH)_2$ with a hexagonal platelet-like shape structure was also observed by XRD and TEM analyses. Thus, the film's final form is believed to have a layered double hydroxides structure, which has a highly beneficial effect on OER activity. SEM analysis showed that the Mn-Co films morphology on nickel foam was characterised by an interconnected 3D nanoflake structure with high porosity.

The Mn-Co films showed promising activity for OER in an alkaline medium (1 M KOH). Their electrocatalytic activity depended on the Mn/Co concentration ratio in the deposition solution and the deposition charge. The film obtained for Mn/Co ratio of 2 mM/8 mM and deposition time limited by a charge of 200 mC showed optimum physicochemical features (e.g., a high specific surface area of 10.5 $m^2g^{-1}$) and optimum electrochemical performance for OER. Thus, it exhibited high electrochemical stability with a low overpotential deviation between 330 and 340 mV at 10 $mAcm^{-2}$ during 70 h.

Supercapacitors are high-performance devices with a high rate of charging/discharging and stability, i.e., capable of effective energy utilisation. As such, they are finding applications in elevators, forklifts, trucks, and buses, i.e., applications demanding substantial amounts of energy to be stored and released in a short time. However, energy densities provided by current supercapacitors are still not high enough to meet all the demands of the modern markets, limiting their usage. Supercapacitors are categorised into three main groups: pseudo-capacitors, electrochemical double-layer capacitors (EDLCs) and hybrid capacitors. Pseudo-capacitors involve a fast Faradaic mechanism (e.g., intercalation or redox reaction), unlike EDLCs that do not include Faradaic processes but the accumulation of ions induced by the adsorption of charged species at the electrode/electrolyte interface.

Supercapacitors comprise two porous electrodes separated by an ionically conducting electrolyte. Different materials, including polymers, carbon and metal oxides, can be used in capacitor electrodes. This special issue presents novel capacitor materials, undoped and Li-ion doped NiO, prepared by Bhatt et al. using a simple microwave method [4]. This approach resulted in the formation of crystalline nanomaterials with a cubic structure. X-ray diffraction analysis also revealed crystallite size decrease upon doping with Li-ion dopant. Furthermore, band gap decrease (3.3 eV for NiO vs. 3.17 eV for NiO doped with 1% Li) and an ultraviolet-blue emission along with a small amount of green emission were observed upon doping revealing the possibility of tuning the nanostructured NiO photoluminescence by doping. The optical results were confirmed and complemented by computational modelling. When it comes to electrochemical behaviour, the maximum reversibility during cyclic voltammetry studies was obtained for a NiO sample with 1% Li doping. The same sample further showed improved electrochemical properties, i.e., conductivity with a reduced charge transfer resistance of $5.592 \times 10^{-8}$ Ω measured by electrochemical impedance spectroscopy, suggesting its potential application in supercapacitors.

Nofal et al. focused their research on electrolytes for supercapacitor applications, specifically on biopolymer-based electrolyte systems comprising methylcellulose (MC) as host polymer material and potassium iodide (KI) as the ionic source [5]. The electrolyte was prepared by a solution cast method using different amounts of KI. The complexation between MC polymer and KI salt was evident from Fourier-transformed infrared spectroscopy analysis. Electrochemical studies were carried out to identify the electrolyte with the highest conductivity for electrochemical double-layer capacitor applications. Increasing the KI concentration from 10 wt.% to 40 wt.% led to an increase of the charge carrier density and, consequently, to a decrease of the resistance ($R_b$) of the charge transfer at the bulk electrolyte by three orders of magnitude (i.e., from $3.3 \times 10^5$ to $8 \times 10^2$ Ω) as determined by the electrochemical impedance spectroscopy measurements. The ionic conductivity was determined to be $1.93 \times 10^{-5}$ $Scm^{-1}$ and dielectric analysis confirmed the conductivity trends. A transference number of 0.88 indicated ions as the dominant charge carriers in the MC-KI electrolyte. The most conducting sample exhibited a wide electrochemical stability window up to 1.8 V during linear scan voltammetry study, pointing out its suitability as an electrolyte for EDLC application. Cyclic voltammetry with activated carbon electrodes displayed an absence of redox peaks and indicated the presence of a charged double-layer between the electrodes' surface and electrolyte. A relatively high value of maximum specific capacitance, $C_s$, of 113.4 F $g^{-1}$ was determined at a polarisation rate of 10 mV $s^{-1}$.

The guest editors believe that this special issue brings valuable guidelines for researchers in the area of electrochemical energy conversion/storage. Finally, we would like to express our gratitude to all authors for their valuable contributions to this special issue, as well as to all reviewers for their time and help in further improving the submitted papers.

**Funding:** Fundação para a Ciência e a Tecnologia (FCT, Portugal) is acknowledged for research contract in the scope of programmatic funding UIDP/04540/2020 (D.M.F. Santos) and contract No. IST-ID/156-2018 (B. Šljukić).

**Conflicts of Interest:** The authors declare no conflict of interest.

## References

1. Hashem, A.M.; Abdel-Ghany, A.E.; Scheuermann, M.; Indris, S.; Ehrenberg, H.; Mauger, A.; Julien, C.M. Doped Nanoscale NMC333 as Cathode Materials for Li-Ion Batteries. *Materials* **2019**, *12*, 2899. [CrossRef] [PubMed]
2. Ajpi, C.; Leiva, N.; Vargas, M.; Lundblad, A.; Lindbergh, G.; Cabrera, S. Synthesis and Characterization of $LiFePO_4$–PANI Hybrid Material as Cathode for Lithium-Ion Batteries. *Materials* **2020**, *13*, 2834. [CrossRef]
3. Cysewska, K.; Rybarczyk, M.K.; Cempura, G.; Karczewski, J.; Łapiński, M.; Jasinski, P.; Molin, S. The Influence of the Electrodeposition Parameters on the Properties of Mn-Co-Based Nanofilms as Anode Materials for Alkaline Electrolysers. *Materials* **2020**, *13*, 2662. [CrossRef]
4. Bhatt, A.S.; Ranjitha, R.; Santosh, M.S.; Ravikumar, C.R.; Prashantha, S.C.; Maphanga, R.R.; Silva, G.F.B.L.E. Optical and Electrochemical Applications of Li-Doped NiO Nanostructures Synthesized via Facile Microwave Technique. *Materials* **2020**, *13*, 2961. [CrossRef] [PubMed]
5. Nofal, M.M.; Hadi, J.M.; Aziz, S.B.; Brza, M.A.; Asnawi, A.S.F.M.; Dannoun, E.M.A.; Abdullah, A.M.; Kadir, M.F.Z. A Study of Methylcellulose Based Polymer Electrolyte Impregnated with Potassium Ion Conducting Carrier: Impedance, EEC Modeling, FTIR, Dielectric, and Device Characteristics. *Materials* **2021**, *14*, 4859. [CrossRef] [PubMed]

*Article*

# Doped Nanoscale NMC333 as Cathode Materials for Li-Ion Batteries

**Ahmed M. Hashem [1,2], Ashraf E. Abdel-Ghany [1], Marco Scheuermann [2], Sylvio Indris [2], Helmut Ehrenberg [2], Alain Mauger [3] and Christian M. Julien [3,*]**

1 National Research Centre, Inorganic Chemistry Department, 33 El Bohouth St., (former El Tahrir St.), Dokki-Giza 12622, Egypt; ahmedh242@yahoo.com (A.M.H.); achraf_28@yahoo.com (A.E.A.-G.)

2 Karlsruhe Institute of Technology (KIT), Institute for Applied Materials (IAM), P.O. Box 3640, 76021 Karlsruhe, Germany; macolino@gmx.de (M.S.); sylvio.indris@kit.edu (S.I.); helmut.ehrenberg@kit.edu (H.E.)

3 Institut de Minéralogie, Physique des Matériaux et de Cosmochimie (IMPMC), Sorbonne Université, Campus Pierre et Marie Curie, UMR 7590, 4 place Jussieu, 75252 Paris, France; alain.mauger@upmc.fr

* Correspondence: christian.julien@upmc.fr

Received: 10 July 2019; Accepted: 5 September 2019; Published: 7 September 2019

**Abstract:** A series of $Li(Ni_{1/3}Mn_{1/3}Co_{1/3})_{1-x}M_xO_2$ ($M$ = Al, Mg, Zn, and Fe, $x$ = 0.06) was prepared via sol-gel method assisted by ethylene diamine tetra acetic acid as a chelating agent. A typical hexagonal α-$NaFeO_2$ structure ($R$-3$m$ space group) was observed for parent and doped samples as revealed by X-ray diffraction patterns. For all samples, hexagonally shaped nanoparticles were observed by scanning electron microscopy and transmission electron microscopy. The local structure was characterized by infrared, Raman, and Mössbauer spectroscopy and $^7$Li nuclear magnetic resonance (Li-NMR). Cyclic voltammetry and galvanostatic charge-discharge tests showed that Mg and Al doping improved the electrochemical performance of $LiNi_{1/3}Mn_{1/3}Co_{1/3}O_2$ in terms of specific capacities and cyclability. In addition, while Al doping increases the initial capacity, Mg doping is the best choice as it improves cyclability for reasons discussed in this work.

**Keywords:** sol-gel synthesis; EDTA chelator; cathode materials; layered oxide; doping; lithium-ion batteries

## 1. Introduction

Lithium-ion batteries (LIBs) applied to power different systems such as portable electronics and electric cars require high-power density, fast charge/discharge rates, and long cycling lives [1–3]. It is well known that the positive electrode (cathode) in LIBs plays an essential role in guiding the electrochemical performance and is considered as the limiting element of the battery. Used in the first generation of LIBs, the layered oxide $LiCoO_2$ (LCO) crystallizing with the α-$NaFeO_2$ type structure [4] demonstrated good reversibility and high-rate capability making this material very popular in commercial lithium-ion batteries. However, some drawbacks have been identified including high cost, toxicity, limited practical capacity (~130 mAh $g^{-1}$; $\Delta x(Li) \approx 0.5$ in the voltage range 2.7–4.2 V), poor thermal stability, and fatigue on crystal structure, which will limit further use of LCO in the forthcoming years [5–7].

Intensive studies for the replacement of LCO were conducted by formulating layered oxides with multi-components of the system $LiMO_2$ ($M$ = Ni, Mn, and Co); then, the new lamellar compounds $LiNi_{1-y}Mn_yO_2$ (NMO), $LiNi_{1-y}Co_yO_2$ (NCO), and $LiNi_yMn_zCo_{1-y-z}O_2$ (NMC) were successively proposed [8–13]. The performance of $LiNi_xMn_yCo_{1-x-y}O_2$ as cathode material relies on the distribution of the transition-metal (TM) cations. For $y = z$, the Ni and Mn ions are, respectively, in 2+ and 4+ oxidation state. $Ni^{2+}$ ions ($r_{(Ni2+)}$ = 0.69 Å) are active species, while $Mn^{4+}$ ions ($r_{(Mn4+)}$ = 0.54 Å) stabilize the structure of the α-$NaFeO_2$ phase [14]. However, high Mn content in the NMC composition may cause phase transition from the layered to the spinel structure [15]. $Co^{3+}$ ions ($r_{(Co3+)}$ = 0.545 Å) play a role in reducing the cation mixing, corresponding to the anti-site defect where $Ni^{2+}$ and $Li^+$ ions exchange their site (interlayer mixing). This defect results from the fact that the ionic radius of $Li^+$ (0.74 Å) is close to that of $Ni^{2+}$ [16,17]. Therefore, the content of transition-metal cations in $LiNi_xMn_yCo_{1-x-y}O_2$ should be optimized to give the best performance as an active cathode material in lithium batteries. In this framework, the composition $LiNi_{1/3}Mn_{1/3}Co_{1/3}O_2$ (named NMC333 hereafter) is of particular interest, as it delivers a reversible capacity of approximately 200 mAh $g^{-1}$ in the voltage range 2.8–4.6 V versus $Li^+$/Li [18,19]. Despite the advantage of high capacity, these materials display the shortcoming of capacity fading when cycled at high voltage and high-rate current density [19]. Two effective approaches have been successfully used to improve their electrochemical performance: Addition of substituting element (doping) and surface modification (coating) [20,21]. Fergus [22] reviewed the effect of doping on the electrochemical performance of cathode materials pointing out that the analysis of doping effects is complicated by the dopant–host structure interrelations modifying the microstructure and morphology. Appropriate metal doping will improve the structural integrity of the oxide framework and hinders oxygen release, because of the higher *M*–O bond dissociation energy in the $MO_6$ octahedron [23].

Many attempts have been made to insert TM ions in NMC frameworks using various elements such as other TM ions (Ti, Zr, Nb, Fe, Cr, Cu) [24–30], rare earths (La, Ce, Pr, Y) [31–34], post-transition metal (Zn) [35], and *sp* elements (Al, Mg), which are the most popular dopants [36–39]. Depending on the nature of the substituting atom, the crystal structure and electrochemical performance of the electrode are differently modified. Substitution of Al and Mg for TMs in NMCs enhances the thermal stability and improves the electrochemical stability of this cathode material [40–42]. Mg doping has improved NMC333 electrodes by modifying the microstructure and reducing the charge transfer resistance [29]. Studies of the impact of Co-selective substitution by Ti, Al, and Fe showed that ~8% $Ti^{4+}$ improves the rate capability, ~5% $Al^{3+}$ improves the capacity retention, while $Fe^{3+}$ doping is detrimental to the electrochemical performance due to the increase of concentration of anti-site defects (implying a *c/a* ratio reduction) resulting in kinetic limitations in NMC333 [28,36]. Aluminum is a very commonly used substituting element in NMC333 cathode materials [28,29,43–45]. Contrary to the expectation of a decreased capacity as $Al^{3+}$ ions are not active in the considered electrochemical window (cannot show further oxidation), one observes better performance due to the improved electrode kinetics, structural modifications, and microstructural effects. However, it must be pointed out that the major differences compared to other reported studies come from the different preparation methods producing particles with various morphologies in connection with the shape of grains, particle size, specific surface area, and particle size distribution. Numerous works have shown that wet-chemistry is a powerful technique to optimize morphology of oxides; thus, using ethylene diamine tetra acetic acid (EDTA) as the chelating agent during the sol-gel synthesis may be unique in this study.

In this work, we investigated the effect of a partial substitution of the TMs on structural, morphological, and electrochemical properties of a series of $Li(Ni_{1/3}Mn_{1/3}Co_{1/3})_{0.94}M_{0.06}O_2$ cathode materials with divalent $Zn^{2+}$, $Mg^{2+}$, and trivalent $Fe^{3+}$ and $Al^{3+}$ ions. The samples were synthesized using a sol-gel method assisted by EDTA as chelator. Ethylene diamine tetra acetic acid, which can chelate several metal ions at the same time, has a unique property of reducing the calcination temperature for the oxide preparation. Recently, we reported the efficiency of this method to prepare nanostructured $LiMn_2O_4$ powders with well-controlled particle size and size distribution [46]. The reason is that EDTA acts as a hexadentate ligand including six sites (i.e., two amines and four carboxyl groups) that can bind to the metal ions. The EDTA forms stable and strong complexes with metal ions through strong masking of free metal ions, which alleviates negative effects such as precipitation of sparingly soluble salts. The strong chelating power of EDTA makes possible the accurate control of dopant concentrations with accuracy better than 1%. Attention has been paid to synthesize all the samples with almost the same morphology and particle size, so that direct comparison of the electrochemical properties of NMC333 cathodes doped with Al, Mg, Fe and Zn is meaningful.

## 2. Materials and Methods

$Li(Ni_{1/3}Co_{1/3}Mn_{1/3})_{1-x}M_xO_2$ ($M$ = Fe, Al, Mg and Zn, $x$ = 0.06) materials were prepared by a sol-gel method EDTA as a chelating agent. Li, Ni, Mn, Co, Mg, Zn, and Al acetates and iron citrate (Merck KGaA, Darmstadt, Germany, 99.99% grade) were used as starting materials. Proper amounts of these starting materials, according to the desired stoichiometry, were dissolved in de-ionized water to form aqueous solution. The dissolved solutions were added stepwise into a stirred aqueous solution of EDTA with a 1:1 metal/chelator ratio. The solution was stirred for 3 h for a homogenous mixture of the reaction reagents and favor complex reaction between metal ions and EDTA. Ammonium hydroxide was added to adjust pH of the solution at ~7. Transparent gel was formed after slow evaporation of the solution. The resulting precursor was heated and decomposed at 450 °C for 5 h in air then ground and recalcined at 700 °C for 8 h in air.

The crystal structure of the prepared materials was investigated by X-ray diffraction (XRD) using Philips X'Pert apparatus (Hamburg, Germany) equipped with CuKα X-ray source (λ = 1.5406 Å). X-ray diffraction measurements were collected in the 2θ range 10–80° at low scanning rate. Sample morphology was observed by scanning electron microscopy (SEM; JEOL model JEM-1230, Freising, Germany). Fourier transform infrared (FTIR) spectra were recorded with a vacuum interferometer (model IFS 113 (Bruker, Karlsruhe, Germany). In the far-infrared region (800–100 $cm^{-1}$), the vacuum bench apparatus was equipped with a 3.5 μm-thick Mylar beam splitter, a globar source, and a DTGS/PE far-infrared detector. Raman scattering (RS) spectra were measured using a micro-Raman-laser spectrometer model Lab-Ram (Horiba-Jobin-Yvon, Longjumeau, France) equipped with 50× microscope lens, D2 filter, aperture of 400 μm, and a slit of 150 μm. The spectra have been recorded with the red (λ = 632 nm) laser excitation. $^{57}$Fe Mössbauer spectroscopic measurements were performed in transmission mode at room temperature using a constant acceleration spectrometer with a $^{57}$Co (Rh) source. Isomer shifts are given relative to that of α-Fe at room temperature. $^7$Li magic-angle spinning (MAS) NMR was performed on a Bruker Avance 200 MHz spectrometer ($B_0$ = 4.7 T) using 1.3 mm zirconia rotors in a dry nitrogen atmosphere. An aqueous 1 mol $L^{-1}$ LiCl solution was used as the reference for the chemical shift of $^7$Li (0 ppm). $^7$Li one-dimensional magic angle spinning nuclear magnetic resonance (MAS NMR) experiments were performed at 298 K and a spinning speed of approximately 60 kHz with a rotor synchronized Hahn-echo sequence ($\pi$/2–τ–$\pi$–τ– acquisition). The typical values for the recycle delay and the $\pi$/2 pulse length were 1 s and 2 μs, respectively.

Electrochemical tests were performed using a multichannel potentiostatic-galvanostatic system VMP3 (Biologic, Grenoble, France). The cathode mixture for the fabrication of the positive electrode was prepared by mixing 80 wt.% of the active material with 10 wt.% of super P® Li carbon (TIMCAL) and 10 wt.% of polyvinylidene fluoride binder (PVDF), dissolved in N-methylpyrrolidone (NMP). The electrode was formed by loading this mixture at 6 mg $cm^{-2}$ onto an Al foil dried at 100 °C for 1 h in a vacuum oven. Coin cells were assembled in an argon-filled dry box with Li foil as anode and glass-fiber separator soaked with 1 mol $L^{-1}$ $LiPF_6$ in ethylene carbonate/dimethyl carbonate (EC/DMC) (1:1 by v/v) aprotic solution. Galvanostatic charge-discharge cycling was carried out at C/10 rate in the voltage range 2.5–4.5 V versus $Li^+/Li^0$.

## 3. Results and Discussion

### 3.1. Structural Analysis

The XRD diagrams of pristine $LiNi_{1/3}Mn_{1/3}Co_{1/3}O_2$ and doped $Li(Ni_{1/3}Mn_{1/3}Co_{1/3})_{1-x}M_xO_2$ powders ($M$ = Fe, Al, Mg, Zn; $x$ = 0.06) are presented in Figure 1a. We can clearly observe well-separated XRD reflections for all prepared samples with very smooth background, indicating highly crystallized products. These diffraction peaks are indexed to a hexagonal lattice α-$NaFeO_2$-type

(*R-3m* space group, standard card JCPDS 82-1495) without any impurity phase, reflecting the formation of a single-phase layered structure. This unique and well-developed layered structure can be confirmed from the distinct splitting of the 108/110 and 006/102 doublets. Values of the lattice parameter *a* (which related to the average metal-metal intraslab distance) and *c* (which related to the average interslab distance) calculated for parent and doped NMC333 are summarized in Table 1. The lattice parameters of parent $LiNi_{1/3}Mn_{1/3}Co_{1/3}O_2$ were consistent with previous reported results [36,47]. The material being an ionic compound, the dopant ions were inserted preferentially on sites that minimize the cost in Coulomb energy. As a first consequence, none of the dopants were expected to substitute for $Mn^{4+}$. Mg and Zn being divalent were substituted for $Ni^{2+}$. As we shall see later in this work, the Mössbauer results showed that iron was introduced in the matrix in the low-spin $Fe^{3+}$ state. Therefore, according to the rule mentioned above, both $Al^{3+}$ and $Fe^{3+}$ substitute for $Co^{3+}$. The Rietveld refinements based on the XRD patterns were done accordingly, including the possibility for the dopants to create also an anti-site defect. Indeed, an important character of these layered rock–salt structures is the cation mixing, occupation of the 3*b* Li interlayer sites by $Ni^{2+}$ ions in the pristine NMC333, and possibly also by dopant metal ions. The results of the refinement are displayed in Figure 1b–f and data listed in Table 1. The residual and agreement parameters (*R* and $\chi^2$) of the Rietveld refinement were very good using this model, taking into account such a cation mixing.

By introducing $Mg^{2+}$ ions in the structure, the lattice expansion occurred in both the *a* and *c* directions, due to the slightly larger radius of $Mg^{2+}$ ($r_{(Mg2+)}$ = 0.72 Å) compared to that of $Ni^{2+}$ ($r_{(Ni2+)}$ = 0.69 Å). Reciprocally, the ionic radius of $Al^{3+}$ ($r_{(Al3+)}$ = 0.53 Å) was close to but a little bit smaller than that of low-spin $Co^{3+}$ ($r_{(Co3+)}$ = 0.545 Å), so that the substitution mainly induces a small shrinking of the lattice parameters. In both cases, however, the lattice distortion is small since the ionic radii are close. On the contrary, the radius of $Fe^{3+}$ ($r_{(Fe3+)}$ = 0.645 Å) was much larger than that of $Co^{3+}$, even larger than high-spin $Co^{3+}$. Nevertheless, there is a general agreement for the introduction of $Fe^{3+}$ ($r_{(Fe3+)}$ = 0.645 Å) on $Co^{3+}$ positions [23,28,36,48], so that the rule mentioned above remains valid, despite the strong lattice distortion that this substitution will cost. The substitution of $Fe^{3+}$ for $Co^{3+}$ leads a lattice expansion, i.e., 0.4% and 0.2% for the *a*- and *c*-axis, respectively. On another hand, the *c*/*a* ratio was reduced, which suggests an increase in the anti-site cation defect concentration [28]. In the same way, a large lattice distortion was expected with the substitution of $Zn^{2+}$ for $Ni^{2+}$, since the size of $Zn^{2+}$ ($r_{(Zn2+)}$ = 0.74 Å) was bigger than that of $Ni^{2+}$. One then would expect an expansion of the unit-cell parameters as in the case of $Mg^{2+}$ and $Fe^{3+}$. However, for $Zn^{2+}$ doping, the data in Table 1 show that the *a*-parameter slightly increased, while the *c*-lattice parameter was not affected. This abnormal behavior provides evidence of important lattice distortions associated to the incorporation of $Zn^{2+}$. It is also the indication of the degree of cation mixing in 3*a* and/or 3*b* sites, in agreement with Table 1.

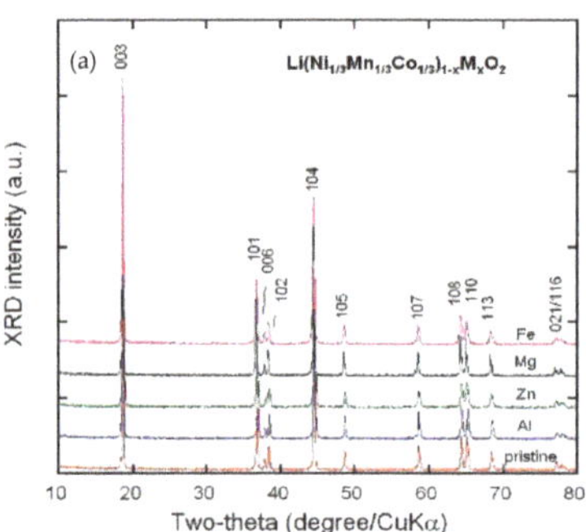


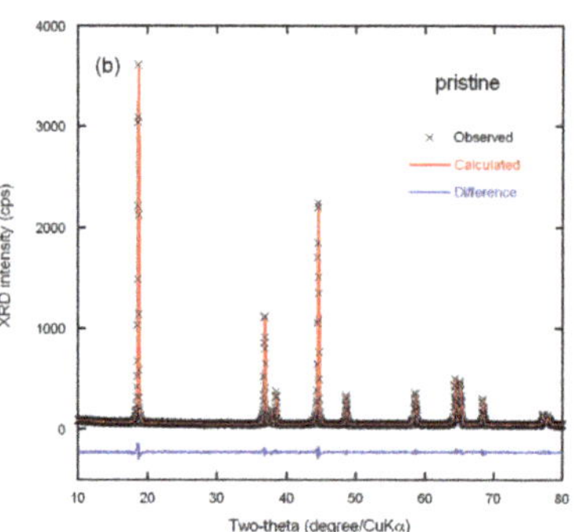


**Figure 1.** *Cont.*

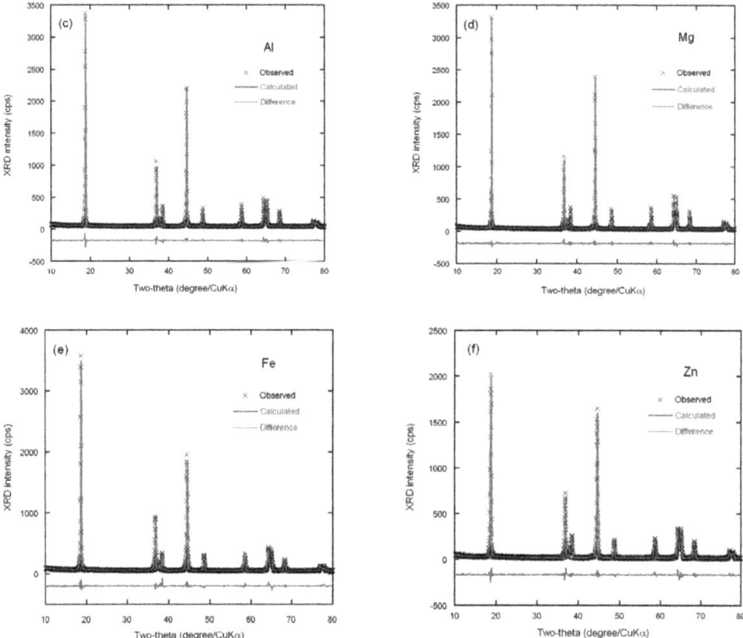


**Figure 1.** X-ray diffraction (XRD) patterns (**a**) and Rietveld refinements (**b–f**) of undoped $Li(Ni_{1/3}Mn_{1/3}Co_{1/3})O_2$ and doped $Li(Ni_{1/3}Mn_{1/3}Co_{1/3})_{1-x}M_xO_2$ ($x$ = 0.06, $M$ = Fe, Al, Mg, Zn), respectively. Cross marks are experimental data and solid lines (in red) are calculated diagrams. The curve at the bottom (in blue) is the difference between the calculated and observed intensities.

**Table 1.** Results of Rietveld refinements for undoped $Li(Ni_{1/3}Mn_{1/3}Co_{1/3})O_2$ and doped $Li(Ni_{1/3}Mn_{1/3}Co_{1/3})_{1-x}M_xO_2$ ($x$ = 0.06) ($M$ = Fe, Al, Mg, Zn).

| Crystal Data | Parent | Doping Element (*M*) | | | |
|---|---|---|---|---|---|
| | | **Mg** | **Al** | **Fe** | **Zn** |
| Lattice parameters | - | - | - | - | - |
| $a$ (Å) | 2.856 (2) | 2.865 (4) | 2.858 (5) | 2.867 (1) | 2.864 (7) |
| $c$ (Å) | 14.249 (5) | 14.259 (7) | 14.255 (9) | 14.276 (7) | 14.243 (1) |
| $c/a$ | 4.980 (1) | 4.976 (3) | 4.987 (1) | 4.979 (4) | 4.964 (9) |
| $V$ (Å$^3$) | 100.65 | 101.39 | 100.88 | 101.62 | 101.22 |
| $L_c$ (nm) | 18.7 | 17.3 | 25.6 | 14.8 | 15.8 |
| $<e^2> \times 10^{-5}$ (rd$^2$) | 0.2899 | 0.4221 | 0.1303 | 0.8743 | 0.8458 |
| $I_{(003)}/I_{(104)}$ | 1.61 | 1.83 | 1.53 | 1.49 | 1.22 |
| $(I_{(006)} + I_{(102)})/I_{(101)}$ | 0.47 | 0.42 | 0.49 | 0.58 | 0.56 |
| Residuals | - | - | - | - | - |
| $R_p(\%)$ | 8.17 | 8.44 | 8.92 | 9.15 | 10.21 |
| $R_{wp}(\%)$ | 9.01 | 9.69 | 11.07 | 11.50 | 13.81 |
| $R_F$ | 1.68 | 2.16 | 3.16 | 5.31 | 4.12 |
| Occupancy (Occ) | - | - | - | - | - |
| $Ni^{2+}$ on Li-site | 0.0241 | 0.0024 | 0.0142 | 0.0129 | 0.0113 |
| $M$ on Li-site | - | 0.0127 | - | - | 0.0222 |
| (0,0,z) for $O_2$ | 0.25921 | 0.2607 | 0.25964 | 0.2585 | 0.25230 |
| $S(MO_2)^a$ (Å) | 2.1139 | 2.0714 | 2.1011 | 2.1367 | 2.2882 |
| $I(LiO_2)^b$ (Å) | 2.6393 | 2.6817 | 2.6508 | 2.6222 | 2.4595 |

[a] $S(MO_2) = 2((1/3) - Z_{oxy})c$ is the thickness of the metal–$O_2$ planes; [b] $I(LiO_2) = c/3 - S(MO_2)$ is the thickness of the inter-slab space.

It should be pointed out that the increase in the degree of cation mixing is known to limit kinetics that cause poor rate capability of the Fe-doped electrode. However, the ionic radius ($r_{(Mg2+)}$ = 0.72 Å) of $Mg^{2+}$ being close to that of $Li^+$ ($r_{(Li+)}$ = 0.76 Å), the anti-site defect corresponding to a local exchange Mg–Li is likely, i.e., a small fraction of the $Mg^{2+}$ ions will occupy the Li-sites.

The cation mixing can be detected by different ratios. First, the *c/a* lattice ratio indicates the deviation of the rock–salt structure (i.e., *c/a* > 4.90). Higher value of peak ratio $I_{003}/I_{104}$ is the second indicator for a lower amount of undesirable cation mixing and better hexagonal structure [49]. The *R* factor ($R = (I_{006} + I_{102})/I_{101}$) is the third fingerprint of the hexagonal ordering; the lower the *R*, the better the hexagonal ordering [50,51]. However, note that rules concerning $I_{003}/I_{104}$ and *R* are well established for undoped NMC samples. The distortion of the lattices associated to the doping may also affect these parameters. For instance, we see in Table 1 that, according to the more reliable Rietveld refinement results, samples doped with Mg and Al have lower cation mixing with well-ordered rhombohedral structures, despite the fact that $I_{003}/I_{104}$ in the Al-doped sample was smaller than in the pristine sample.

The doping by any of the elements investigated reduces the concentration of $Ni^{2+}$–Li anti-site defects. In particular, the introduction of Mg ions on *3a* sites led to a remarkable decrease of $Ni^{2+}$ ions on the *3b* site (Table 1); this observation matches well with previous reports [52,53]. However, while most of the Mg, Fe, and Zn ions were the *3a* sites, the Rietveld refinement showed that the *3b* Li site was also partly occupied by these foreign ions. This feature is attributable to the fact that the ionic radii of $Mg^{2+}$, $Fe^{3+}$, and $Zn^{2+}$ are comparable to that of $Li^+$ ($r_{(Li+)}$ = 0.76 Å) [54]. On the one hand, the much smaller ionic radius of $Al^{3+}$ can explain the absence of $Al^{3+}$ ions on the lithium site. Overall, the cation mixing obtained by adding the concentration of nickel and dopant ions on the *3b* lithium sites is 1.4% and 1.5% for Al- and Mg-doped samples, respectively, smaller than 2.4% in the undoped sample. On the other hand, this concentration was larger in the Zn- and Fe-doped samples.

Two other structural parameters can be deduced from Rietveld refinements: $I(LiO_2)$ the thickness of the inter-slab space and $S(MO_2)$ the thickness of the metal–$O_2$ planes [55]. As seen from Table 1, both *a*- and *c*-parameters and $S(MO_2)$ were minimum and $I(LiO_2)$ was maximum for *sp*-doped elements, which confirms that these samples had better structural integrity.

The cation mixing between metal ions ($Ni^{2+}$ and $M^{2+\text{ or }3+}$) and Li ions on the two sites resulted in two competitive effects on lattice parameter depending on the ionic radius and ionic charge of metal: (i) As $r(Li^+)$ was larger than the ionic radii of the other Ni, Mn, and Co ions, this difference favored an increase of the in-plane parameter *a* due to the presence of Li ions on the *3a* site. Moreover, the $Li^+$ ions carried only a charge +1, so that the occupation of *3a* sites by $Li^+$ decreased the repulsive Coulomb potential with the neighboring TM ions inside the slabs. This effect also favored an increase of the *a*-lattice parameter, (ii) this effect is in part compensated by the concomitant presence of the $Ni^{2+}$ and $M^{2+\text{ or }3+}$ on the *3b* site. Moreover, the metal ions on the *3b* site carried more charge than $Li^+$, which leads to a stronger electrostatic attraction between these ions and $O^{2-}$ in the interslab plane. As a consequence, the Li-O interslab distance and the related *c*-lattice parameter decreased. This stronger Li–O bond also hinders $Li^+$ diffusion through the NMC framework, so that the $Li^+$ ions on the *3a* sites do not contribute to the electrochemical process [17].

The broadening of reflections is an indicator not only to the crystallinity of the $Li(Ni_{1/3}Mn_{1/3}Co_{1/3})_{1-x}M_xO_2$ powder but also to the local deformation of the structure. The combination of the Scherrer's equation for crystallite size with the Bragg's law for diffraction leads to Equation (1):

$$B^2 \cos^2 \theta = 6\langle e^2 \rangle \sin^2 \theta + \frac{K^2\lambda^2}{L_c} \quad (1)$$

which can be used to determine coherence length $L_c$ and micro-strain field $<e^2>$. *B* is the full-width at half-maximum (FWHM) in radian, θ is the diffraction angle, and *K* is a near-unity constant related to crystallite shape. The first member is reported as a function of $\sin^2\theta$ in Figure 2 for the pristine and doped $Li(Ni_{1/3}Mn_{1/3}Co_{1/3})O_2$ samples. The plots are well fit by straight lines, in agreement with Equation (1). The slope of the lines gives the value of the strain field $<e^2>$, while the coherence length $L_c$ is given by the extrapolation to sin θ = 0. For all investigated samples, values were in the range $15 \leq L_c \leq 25$ nm. On the other hand, we found that $<e^2>$ was strongly affected by the nature of the doping elements. While the strain field was negligible in the undoped and Mg-doped samples, it rose to $<e^2> = 0.4221 \times 10^{-5}$ rd$^2$ in Al-doped samples, and became as large as $0.8743 \times 10^{-5}$ and $0.8458 \times 10^{-5}$ rd$^2$ in Fe- and Zn-doped samples, respectively.

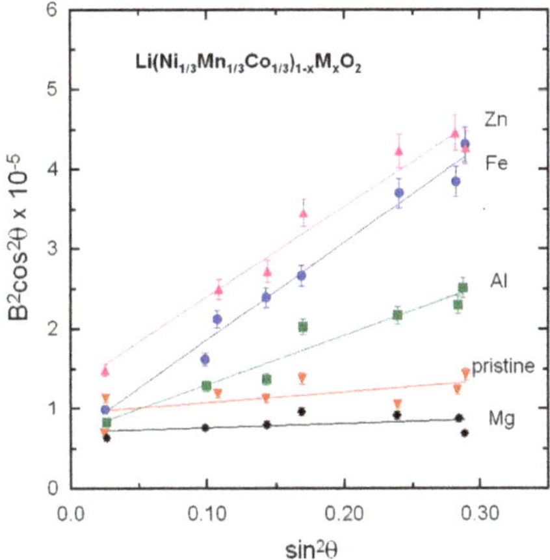


**Figure 2.** Analysis of the full-width at half-maximum, *B*, of XRD peaks according to Equation (1). *B* is expressed in radian.

The negligible value of $<e^2>$ in Mg-doped samples proves that Mg stabilizes the structure of the lattice, a result that is consistent with the fact that it almost totally eliminated the Li-Ni anti-sites. $<e^2>$ is non-negligible in Al-doped samples. This is also consistent with the fact that the concentration of Li-Ni anti-sites, although smaller than in the pristine sample, was not eliminated by the Al doping. This result gives evidence of a better structural stability of the Mg-doped sample than the Al-doped sample. For the Zn- and Fe-doped cases, the large values of $<e^2>$ imply important lattice distortions that will further weaken the structural stability of Fe- and Zn-doped samples, as expected from the discussion reported above on the lattice parameters and the consideration of ionic radii.

### 3.2. Morphology

Figure 3 presents the SEM images of the undoped- and doped-NMC333 materials. The nanoscale powders consisted of hexagonally like shaped particles with flat facets. As the morphology is one of the main factors which affects the electrochemical performance of the electrode materials, we tried to maintain identical aspects for all the samples. However, although the synthesis conditions were the same, the particle size and morphologies depended on the doping element because of the change in the reaction equilibrium despite similar temperature, pH value, etc. The undoped NMC333 showed regular particles, 300–500 nm in size, with quite narrow size distribution (Figure 3a). The Al-doped sample became more uniform and exhibited smaller primary particles with size lying between 200 and 450 nm (Figure 3b). The Mg-substituted sample showed larger particle sizes (Figure 3c). The Fe-doped sample had a broad size distribution with particles in the range 100–600 nm (Figure 3d). The Zn-doped material presented less faceted particles with a narrower size distribution (i.e., 300–500 nm); particles tended to form microspheres (Figure 3e).

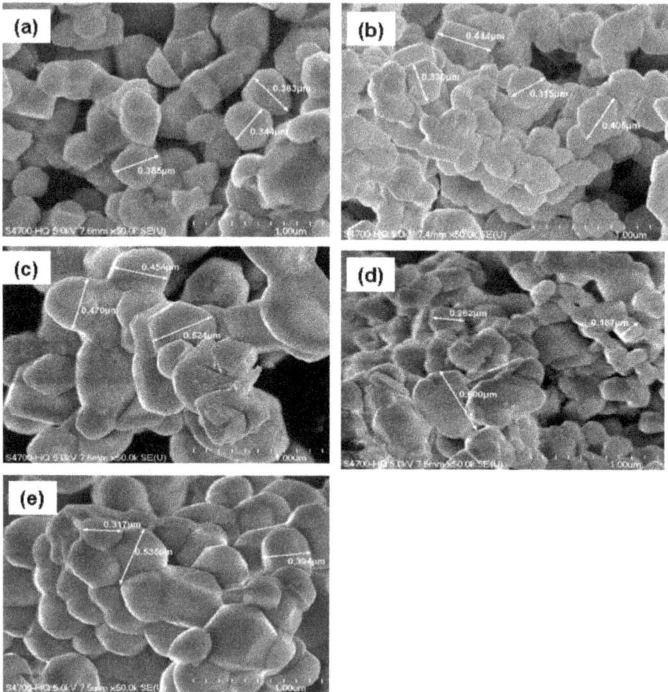

**Figure 3.** Field-emission SEM images of (**a**) undoped and (**b**–**e**) *M*-doped NMC333 powders with *M* = Al, Mg, Fe, Zn, respectively.

All powders maintain the initial morphology of the undoped sample, except a tendency of agglomeration in the case of Fe-doped sample (Figure 3e). It seems that the formation of agglomerates depends strongly of the synthesis technique. Ren et al. [38] reported agglomeration of Al-doped NMC powders even at low concentration of aluminum ($y = 1/12$) when prepared by solvent evaporation method. On another hand, Lin et al. [43] showed well-dispersed particles with a slight decrease of grain size for moderated Al-doped powders ($y = 0.1$) obtained via sol-gel method using polyvinyl alcohol as organic fuel. In conclusion, all these NMC oxides showed almost well-dispersed primary particles with relatively bright and clear surface; in addition, the particles did not display a change in surface roughness with doping. The absence of aggregated particles was attributed to the synthesis process via EDTA chelating assistance with an efficient pH control.

To further examine the sample morphology, Figure 4 presents the typical TEM images of undoped (a) and Al- (b) and Mg-doped NMC333 samples (c) confirming the submicronic size of the particles. The high magnification TEM (HRTEM) micrographs of individual nanoparticle reveal well-defined lattice fringes with a separation of 4.72 Å corresponding to the (003) plane. One also observes that the edges of all as-crystallized particles were well defined (i.e., without disordered surface layer). Therefore, as we shall see later, the difference of morphology among the different samples was small enough to allow for a direct comparison of their electrochemical properties.

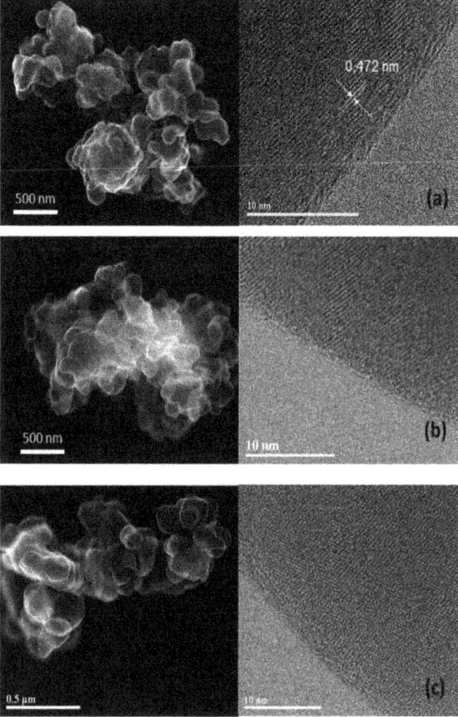


**Figure 4.** TEM images of (**a**) undoped, (**b**) Al-doped, and (**c**) Mg-doped NMC333 samples. The HRTEM micrographs reveal well-defined lattice fringes with a separation of 4.72 Å corresponding to the (003) plane.

### 3.3. Local Structure

The local structure, i.e., short-range environment of lithium within NMC materials was investigated using several analytical methods: Raman, FTIR and Mössbauer spectroscopies, and $^7$Li NMR measurements. Figure 5a,b display the Raman and FTIR spectra of NMC333 samples, respectively. Considering the layered *R*-3*m* structure ($D^5{}_{3d}$ spectroscopic symmetry) for $LiNi_{1/3}Mn_{1/3}Co_{1/3}O_2$, one expects six Raman active modes ($3A_{1g} + 3E_g$) and seven IR active modes ($4A_{2u} + 3E_u$) [56]. As a general trend, the doping did not bring significant alteration in the band positions in both vibrational spectra. Characteristic Raman bands were recorded at 397 and 477 cm$^{-1}$ (O–M–O bending vibrations) and 603 and 641 cm$^{-1}$ (M–O symmetrical stretching), while FTIR patterns were measured at 245 cm$^{-1}$ (Li cage mode), 377 and 472 cm$^{-1}$ (O–M–O asymmetric bending modes), and 527 and 594 cm$^{-1}$ (asymmetric stretching modes of $MO_6$ octahedra). In conclusion, doping did not provoke significant change in spectral features, except a slight broadening of the Li cage mode, which reflects the degree of cationic mixing in the interlayer space.

Solid-state $^7$Li-MAS NMR measurements were carried out to study the structural properties on a local scale, knowing that paramagnetic ions in the surrounding of the lithium ions have strong effects on the NMR spectra due to the Fermi-contact mechanism, i.e., the transfer of spin density from the unpaired electrons of the paramagnetic ions to the lithium nucleus. In $LiNi_{1/3}Mn_{1/3}Co_{1/3}O_2$, $Co^{3+}$ is in its low-spin state and not magnetic, but both $Ni^{2+}$ and $Mn^{4+}$ contribute to the resulting overall hyperfine shifts. Figure 6a shows the $^7$Li-MAS NMR spectra acquired for all NMC333 samples, for which two strong contributions are discernible. A rather narrow resonance was located at around

0 ppm. It can be assigned to diamagnetic impurities such as LiOH or $Li_2CO_3$. The main contribution was a broad group of resonances with large chemical shifts covering the range from 0 to 1000 ppm with a major broad band at approximately 550 ppm. The distribution of the signal intensity among the resonances was sample dependent. The Mg-doped sample showed the smallest 0 ppm peak, which increased in the undoped and the Al-doped sample. The highest 0 ppm intensity was also found for Fe- and Zn-doped NMCs. The broad contributions ranging from 0 to 1000 ppm were analyzed by means of a spectral deconvolution, which revealed the degree of cationic disorder (Table 2). Obviously, the shifts of the three resonances were approximately equidistant. In all samples, the intermediate resonance with a large chemical shift of about 550 ppm, which had the highest intensity, was assigned to the hyperfine interaction between $Li^+$ nuclei and the unpaired electrons of $Ni^{2+}$ and $Mn^{4+}$ paramagnetic ions [57–59]. Note that the broadening of the peak at ~550 ppm suggests the less ordered local environment of Li on 3*b* sites.

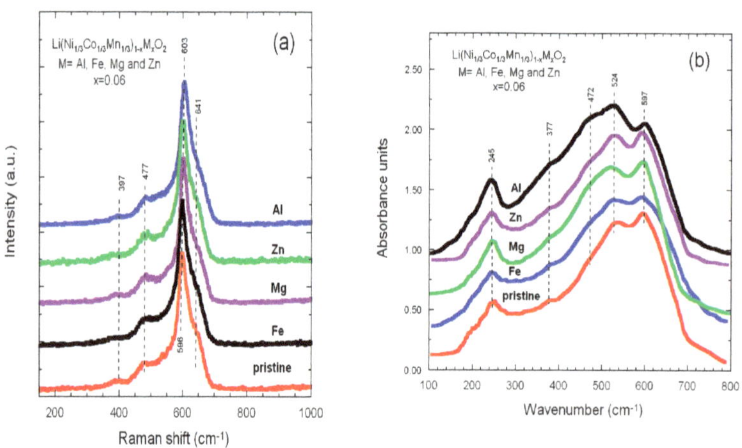


**Figure 5.** (**a**) Raman spectra and (**b**) FTIR of undoped and doped $LiNi_{1/3}Mn_{1/3}Co_{1/3}O_2$ ($x$ = 0.06) powders synthesized via sol-gel method.

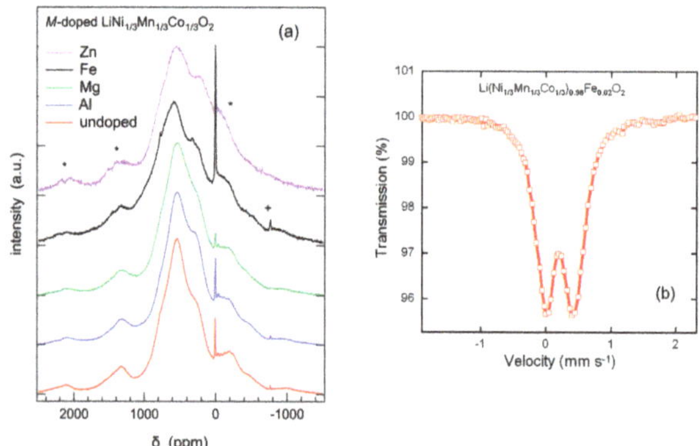


**Figure 6.** (**a**) $^7Li$ MAS NMR spectra of undoped and doped $LiNi_{1/3}Mn_{1/3}Co_{1/3}O_2$ samples; (**b**) Fe Mössbauer spectrum of $Li(Ni_{1/3}Mn_{1/3}Co_{1/3})_{1-x}Fe_xO_2$.

**Table 2.** Deconvolution of sample spectra. Calculated areas include all significant sideband contributions.

| Dopant | Shift (ppm) | $I_{rel}$ | Shift (ppm) | $I_{rel}$ | Shift (ppm) | $I_{rel}$ | %Li/Ni Exchange |
|---|---|---|---|---|---|---|---|
| None | 274.9 | 21.1 | 541.4 | 66.9 | 747.3 | 12.0 | 3.11 |
| Al | 277.1 | 26.2 | 539.8 | 62.0 | 731.0 | 11.7 | 1.65 |
| Mg | 289.0 | 28.5 | 533.4 | 52.1 | 716.6 | 19.4 | 3.03 |
| Fe | 283.5 | 7.0 | 563.4 | 54.9 | 776.1 | 38.1 | 2.08 |
| Zn | 217.9 | 28.8 | 541.0 | 49.7 | 743.0 | 21.5 | 2.74 |

Results shown in Figure 6a match well with those reported by Cahill et al. [58] who determined three resonances in the range from 200 to 800 ppm using $^{6}Li$ NMR. These three NMR bands are assigned to $Li^{+}$ ions in the interlayer slab of the NMC lattice with different TM distributions in the first cation coordination shell. The possible $Ni^{2+}/Co^{3+}/Mn^{4+}$ arrangements exhibiting different chemical shifts are distributed by (a) 1:4:1 at low frequency (<300 ppm), (b) 2:2:2 at intermediate frequency (~550 ppm), and (c) 3:0:3 at high frequency (>700 ppm). Adopting these results, we interpreted the three major resonances identified in the deconvolution of our spectra to stem from lithium ions in the 3*b* site (Li layer). The observed hyperfine shifts can be readily explained by the three different local environments proposed by Cahill et al. [58]. Either one, two or three pairs of $Ni^{2+}$ and $Mn^{4+}$ ions are distributed among the six nearest neighbor positions. According to the nominal stoichiometry, the total number of ions of each type should be the same for all transition metals, leading to symmetric intensity distribution. This expectation is not in accordance with the results from the spectral deconvolution of the samples; only the Zn-doped NMC333 sample might be considered to show roughly a symmetric intensity distribution. Another NMR feature was a minor resonance occurring at around 1300 ppm attributed to $Li^{+}$ ions in 3*a* site of the TM layers as a result of the cation mixing with Ni and Mn in first and second coordination shell. Thus, the Li/Ni exchange rate can be calculated by the ratio of the area of the smaller peak over the larger one including side bands (Table 2) and compared with data from Rietveld refinements. From spectra in Figure 6a, it is obvious that the Li/Ni exchange rate was lower for Al- and Fe-doped samples. There is a general agreement for the cation mixing diminution within the Al-doped $LiNi_{1/3}Mn_{1/3}Co_{1/3}O_2$ framework because Al prevents the presence of Li on the 3*a* site of TM layer and restrains $Ni^{2+}$ in the Li plane [60]. Liu et al. [36] reported that the improved structural stability of NMC333 at low Al doping ($y < 1/20$), whereas Fe doping does not display such behavior even at low Fe content.

The Fe Mössbauer spectrum of $LiNi_{1/3}Mn_{1/3}Co_{1/3}O_2$ doped with Fe recorded at room temperature is shown in Figure 6b. It reveals a narrow doublet with an isomer shift of $0.327 \pm 0.001$ mm $s^{-1}$ and a quadrupole splitting of $0.436 \pm 0.002$ mm $s^{-1}$, which are characteristics of $Fe^{3+}$ ions in high-spin state in octahedral oxygen coordination, as expected for the 3*a* site. No other contributions could be detected in the spectrum.

### 3.4. Electrochemical Properties

The effects of doping on the electrochemical properties were systematically investigated by cyclic voltammetry (CV) and galvanostatic charge-discharge (GCD) measurements in the voltage range of 2.5–4.5 V versus $Li^{+}/Li^{0}$. Figure 7 shows the CV profiles of parent and doped NMC333 oxides recorded at a sweep rate of 0.05 mV $s^{-1}$. The pristine NMC333 electrode displayed a sharp anodic peak (delithiation) at 3.83 V and a cathodic peak (lithiation) at 3.72 V. Within the potential range 2.5–4.5 V, redox peaks are ascribed to the reaction of $Ni^{2+}$ to $Ni^{3+/4+}$, while $Mn^{4+}$ is known to be electrochemically inactive and $Co^{3+/4+}$ takes place at potentials above 4.6 V [19]. Except the pristine sample, all doped NMC333 samples exhibited the same features, i.e., a slight voltage shift of the anodic peak ($\Delta E_{pa}$) between the 1st and 2nd cycle, while the cathodic peak was almost unchanged. $\Delta E_{pa}$ appeared to be 20 mV for Al- and Fe-doped NMC; 90 and 160 mV for Mg- and Zn-doped electrodes, respectively. Similar behavior attributed to surface kinetics was reported by Riley et al. [61] for NMC333 coated with

$Al_2O_3$ by atomic layer deposition. The forthcoming CV cycles show similar redox features, indicating a good reversibility for the lithiation/delithiation process, according to the relation [62]:

$$LiNi_{1/3}Mn_{1/3}Co_{1/3}O_2 \rightleftharpoons Li_{1-x}Ni_{1/3}Mn_{1/3}Co_{1/3}O_2 + xLi^+ + xe^- \quad (2)$$

for which the redox couple $Ni^{2+/4+}$ should be considered in the voltage range 2.5–4.5 V versus $Li^+/Li$. Table 3 summarizes the redox peak potentials ($E_{pa}$, $E_{pc}$, $Ni^{2+/4+}$) of the investigated NMC333 electrodes. The peak potential separation (PPS), i.e., expressed by $E_p = E_{pa} - E_{pc}$, between anodic and cathodic potentials, corresponding to the $Ni^{2+/4+}$ redox process, was measured in the range ($100 \leq E_p \leq 130$ mV), These results compare well with data in the literature [43,61]. Here, we can point out that the particle morphology also plays a crucial role. However, the PPS for the insertion type of materials also significantly depends on the scan rate in CV experiments. Lin et al. [43] reported high $\Delta E_p$ values (≈0.33 V at scan rate of 100 μV $s^{-1}$) for Al-doped NMC synthesized via a sol-gel method. Similarly, Li et al. [63] mentioned $\Delta E_p$ values of about 0.3 V measured at scan rate of 0.1 mV $s^{-1}$ for $LiNi_{1/3}Mn_{1/3}Co_{1/3-x}Al_xO_2$ formed by large secondary particles (agglomerates ~1 μm). Wu et al. [37] reported a depressed $E_p$ value (0.23 V at a scan rate of 0.1 mV $s^{-1}$) for 1%-$Al^{3+}$ substituting $Ni^{2+}$ in NMC333 formed by nanoparticles (100–500 nm size) versus 0.42 V for the undoped sample.

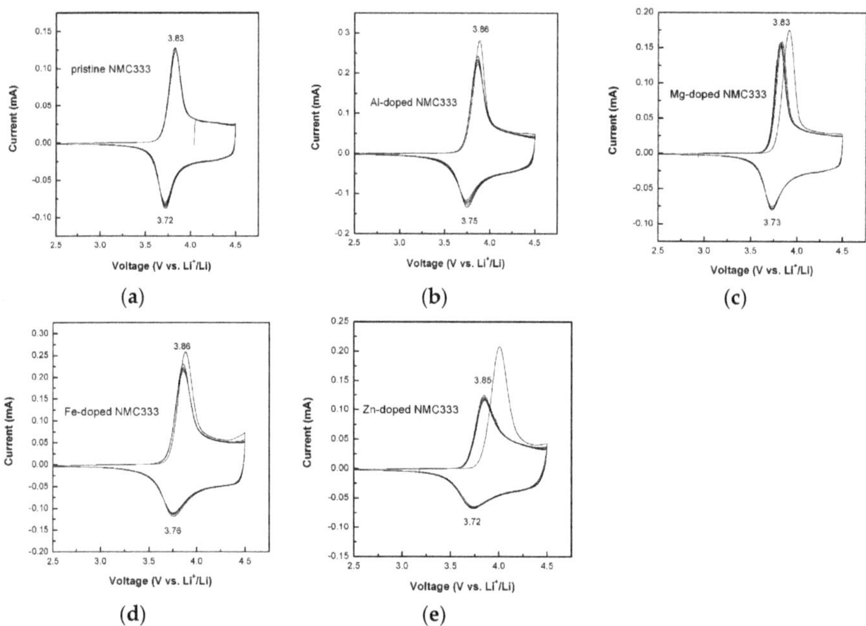


**Figure 7.** Cyclic voltammograms of parent (**a**) and doped $LiNi_{1/3}Mn_{1/3}Co_{1/3}O_2$ with Al (**b**), Mg (**c**), Fe (**d**) and Zn (**e**) at a sweep rate of 0.05 mV $s^{-1}$ between 2.5 and 4.5 V versus $Li^+/Li^0$.

**Table 3.** The oxidation ($E_{pa}$) and reduction ($E_{pc}$) peak potential and the corresponding difference $\Delta E_p$ obtained from CV data (scan rate of 0.05 mV $s^{-1}$) of doped NMC333 cathodes for the 2nd cycle.

| Potential (V Versus $Li^+$/Li) | Doping Element | | | | |
|---|---|---|---|---|---|
| | Pristine | Al | Mg | Fe | Zn |
| $E_{pa}$ | 3.83 | 3.86 | 3.83 | 3.86 | 3.85 |
| $E_{pc}$ | 3.72 | 3.75 | 3.73 | 3.76 | 3.72 |
| $\Delta E_p$ | 0.11 | 0.11 | 0.10 | 0.10 | 0.13 |

Figure 8 presents the galvanostatic charge–discharge curves of Li//NMC333 cells including parent and doped electrodes cycled at a constant current density of 0.1C at 25 °C. Analysis of these results showed that the Al-doped material exhibited better galvanostatic charge/discharge performance. At 0.1C, its initial specific discharge capacity was 160 mAh $g^{-1}$ with a coulombic efficiency of ~85%. No significant change in the initial capacity value upon doping by Mg, similar to bare NMC333, of about 150 mAh $g^{-1}$ was obtained for Mg-doped oxide. Both Fe and Zn doping negatively affect the initial capacities values. Fe-doped oxide delivered 146 mAh $g^{-1}$, whereas Zn-doped material delivers a lower capacity of 118 mAh $g^{-1}$.

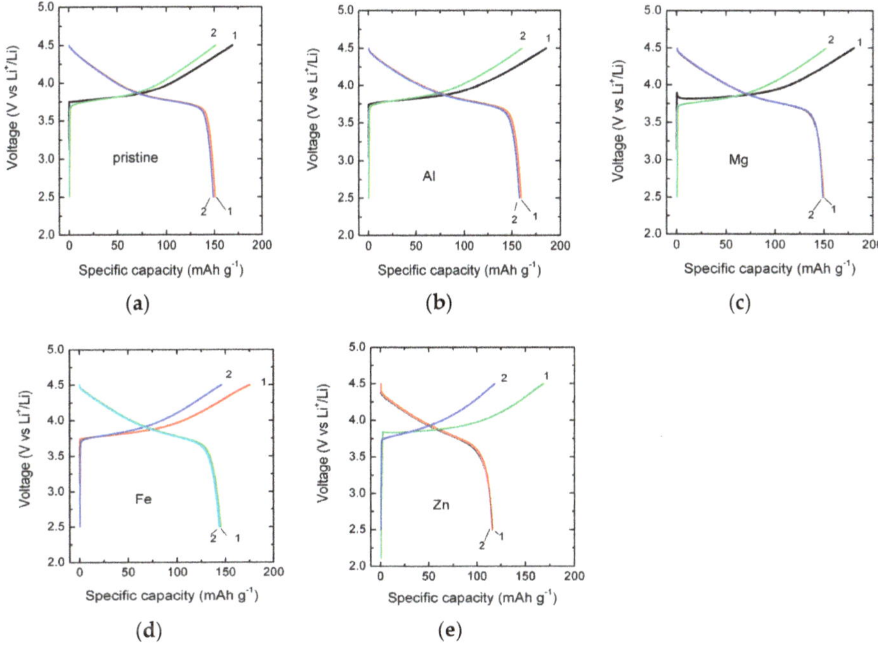


**Figure 8.** Charge-discharge profiles of pristine (**a**) and doped $LiNi_{1/3}Mn_{1/3}Co_{1/3}O_2$ with Al (**b**), Mg (**c**), Fe (**d**) and Zn (**e**) carried out at the 0.1C rate in the potential range 2.5–4.5 V versus $Li^+/Li^0$.

The incremental capacity (IC), i.e., differential capacity ($-dQ/dV$) versus V curve, can be considered as an electrochemical spectroscopy technique [17]. For instance, IC has been successfully applied to analyze the layered and spinel contribution in blended cathodes [64]. The IC curves were extracted from the galvanostatic discharge profiles (lithiation process) during the 2nd cycle to further characterize the electrochemical behavior of doped electrodes, as depicted in Figure 9. Each plot displays a main sharp peak at approximately 3.74–3.78 V versus $Li^+/Li^0$ and a broad voltage peak in the vicinity of 4.25 V, which are typical fingerprints of the $Ni^{2+/3+}$ and $Ni^{3+/4+}$ reactions. These results showed that both $Al^{3+}$ and $Fe^{3+}$ ions were electrochemically inactive for the cutoff charge-discharge voltages in the range 2.5–4.5 V.

Figure 10a compares the cyclability of all samples cycled for 50 cycles at 0.1C. Both Al-and Mg-doped oxides show better rechargeability than bare NMC333, which delivers 85% of its initial capacity after 50 cycles, while Zn and Fe doping display worse results. Doped electrode materials with Mg, Al, Fe, and Zn retained specific capacity of 91%, 82%, 67%, and 36% of the initial values, respectively. Figure 10b presents the electrochemical impedance spectroscopy (EIS) measurements before and after cycling of a cell with Al-doped NMC333 as cathode material. Analysis of the Nyquist plots shows that at high frequency, the electrode/electrolyte resistance, $R_s$ = 14.5 obtained by the intercept of the Z′

axis does not change upon cycling. In contrast, the charge transfer resistance ($R_{ct}$) corresponding to the electrochemical reaction at solid/electrolyte interface (represented by the depressed semicircle at medium frequency region) decreased from 192 for the fresh cell to 165 after 30 cycles. This decrease is attributed to the cell formation occurring after few cycles of charge–discharge. Rate capability was tested in the range from 0.05 C to 5C for the 30th cycle. Results shown in Figure 10c also demonstrated the beneficial effect of Al and Mg doping after 30 cycles of charge-discharge. The specific discharge capacity of pristine NMC333 decreased dramatically down to 59 mAh $g^{-1}$ with increasing current density to 5C, while both capacities of Al- and Mg-doped NMC333 maintained at approximately 112 mAh $g^{-1}$.

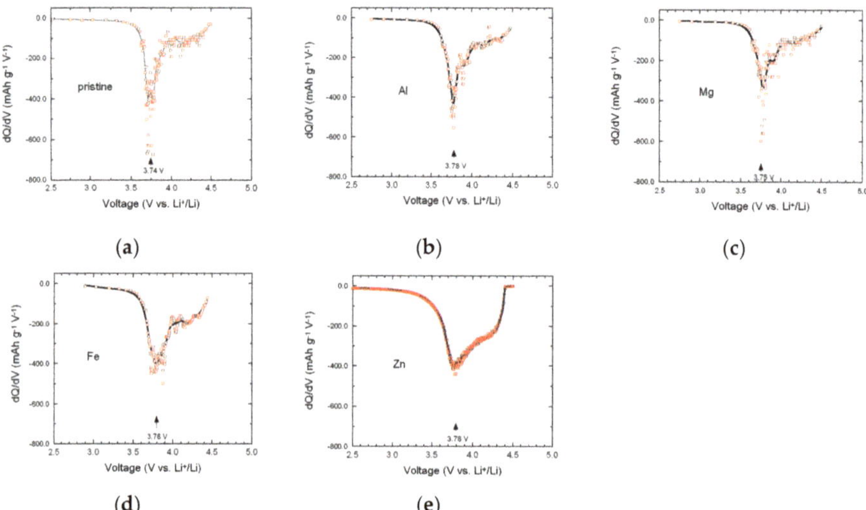


**Figure 9.** Differential capacity ($-dQ/dV$) plots of cycle #2 for pristine $LiNi_{1/3}Mn_{1/3}Co_{1/3}O_2$ (**a**) and *M*-doped $LiNi_{1/3}Mn_{1/3}Co_{1/3}O_2$ with Al (**b**), Mg (**c**), Fe (**d**) and Zn (**e**).

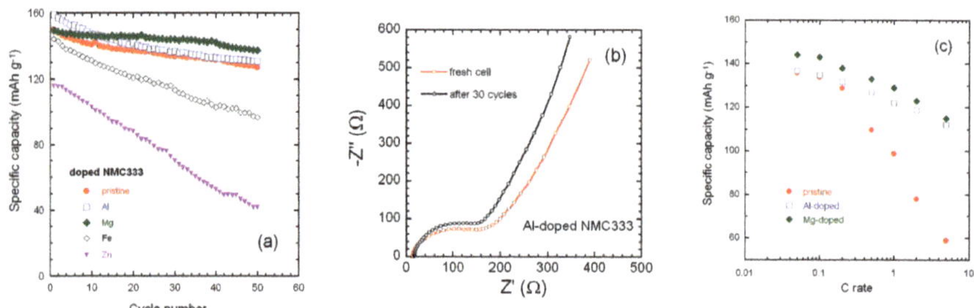


**Figure 10.** (**a**) Cyclability of parent and doped $LiNi_{1/3}Mn_{1/3}Co_{1/3}O_2$ cathode materials recorded at a 0.1C rate in the potential range 2.5–4.5 V versus $Li^+/Li^0$; (**b**) electrochemical impedance spectroscopy (EIS) results for a $LiNi_{1/3}Mn_{1/3}Co_{1/3}O_2$//Li cell before and after cycling at 0.1C rate; (**c**) rate capability of undoped and Al- and Mg-doped NMC333 electrodes after 30 cycles.

## 4. Discussion

In this work, we chose two dopants, Al and Fe, which substituted for Co. Reducing the number of cobalt ions did not penalize per se the capacity when the cell is operating in the voltage range 2.5–4.5 V, because the redox $Co^{3+/4+}$ reaction occurred at a potential >4.6 V. The effects on the electrochemical properties are, however, totally different. The large strain field $<e^2>$ opposed the diffusion of lithium, with the consequence that some of the Li are immobilized so that the initial capacity was smaller than in the pristine sample, and the kinetics were slower, with the consequence that the rate capability was degraded. Moreover, the large strain field gives evidence of a poorer structural stability, implying a rapid decrease of the discharge capacity as a function of the cycle number. To the contrary, Al doping improved significantly the electrochemical properties. The two other dopants, Mg and Zn, substituted mainly on the Ni site and created also anti-site defects, so that a fraction of them also occupy the Li-sites. However, these dopants have also very different impact on the electrochemical properties. Zn doping degrades the electrochemical properties for the same reasons invoked for Fe doping: it provokes a large strain field that immobilizes some of the lithium ions, slows the kinetics, and decreases the structural stability. On another hand, Mg doping improves the electrochemical properties, but differently. The lithium ions in the anti-site defects do not participate in the electrochemical process [17]. Therefore, the more anti-site defects that exist, the smaller the capacity of the battery. The overall concentration of anti-sites is in the order: pristine > Mg-doped > Al-doped, after the results in Table 1. We then expect the initial capacity at a low rate in the opposite order: Pristine < Mg-doped < Al-doped, in agreement with the initial capacity at 0.1C in Figure 10a. The reason is that, even though Mg- oping is efficient to suppress the Li–Ni anti-sites, the Mg–Ni anti-site is inevitable, because $Mg^{2+}$ and $Li^{+}$ have almost the same ionic radius. As a consequence, the overall anti-site concentration in the Mg-doped sample was larger compared to Al-doping, despite the absence of Li–Al anti-sites. On another hand, the significant strain field associated with Al-doping evidenced in Figure 2 and Table 1 was responsible for reduced cycle ability, in agreement with Figure 10a, where the two curves of the capacity versus number of cycles cross each other, so that the capacity was larger in the Mg-doped sample after 12 cycles. The beneficial effect of Mg doping is in good agreement with the result of Luo et al. [65], reporting the beneficial effect of Mg substitution on the degree of cation mixing, although we do not agree with the assumption that Mg substituted for Mn, as claimed in this prior work, as the apparent increase in transport properties, can be attributed to the modification of the microstructure and the slight decrease of the NMC333 particle sizes with Al- and Mg-doping concentrations (> 0.02) [66]. The change in the microstructure was related to the increase of the interlayer distance $I(LiO_2)$ from 2.6393 Å in pristine NMC to 2.6508 and 2.6817 Å in Al- and Mg-doped NMC, respectively (see Table 1). The increased $I(LiO_2)$ results in better mobility of $Li^{+}$ ions in the NMC framework to enhance the rate capability. In contrast, different researchers have shown contradictory results. For example, in Reference [29], the authors reported that Mg dopant in $Li(Ni_{1/3}Co_{1/3}Mn_{1/3})O_2$ cathodes synthesized by hydroxide coprecipitation method did not exhibit improvement and stated an increase of undesirable reactions between the electrode and the electrolyte inducing larger capacity fade. Hence, this lack of improvement seems to be due to the morphology of the NMC powders having very large particle size distribution. Several authors [67,68] proposed that the enhanced cycling stability of Mg substitution samples was attributed to the Mg ions incorporation into interlayer planes due to the similar ionic radii of $Li^{+}$ and $Mg^{2+}$. This is in opposition with the present Rietveld refinements, which determine a small concentration of 1.3% for $Mg^{2+}$ on Li sites and a weak cationic mixing rate of 0.24%.

The improvement of Al-doped electrode relative to bare NMC333 cathode materials is commonly attributed to an enlarged Li layer spacing and a reduced degree of cation mixing [36–39]. Presently, the as-prepared $Li(Ni_{1/3}Mn_{1/3}Co_{1/3})_{0.94}Al_{0.06}O_2$ powders show limited defect in the Li plane, 2.4% $Ni^{2+}$ and 1.3% $Al^{3+}$ on Li site. Note that, in this work, for a relevant comparison, special attention was taken to prepare powders having the same morphology, particle size, and size distribution by adjusting the temperature and duration of the annealing process [69]. Both the charge and discharge voltage plateaus of Al-doped NMC333 were higher than those of the pristine electrode. First-principles calculations also predicted the potential increase with the substitution of Co with Al [70]. This phenomenon, experimentally observed by Julien et al. [71] in Al-doped $LiNi_{0.5}Co_{0.5}O_2$ and by Liu et al. [36] in $LiNi_{1/3}Co_{1/3}Mn_{1/3}O_2$, was due to the drop in the chemical potential of the material. The lowering delithiation potential (end of charge) for Fe-doped $LiNi_{1/3}Co_{1/3}Mn_{1/3}O_2$ was also identified by Meng et al. [40], while the raise in the delithiation voltage for $Al^{3+}$ doping of NMC materials was reported by Liu et al. [36]. Our results are in good agreement with those reported by Samarasingha et al [72] on $Li(Ni_{1/3}Mn_{1/3}Co_{(1/3-x)}Fe_x)O_2$ with $x = 0.11$ for which an initial specific discharge capacity of ca. 126 mAh $g^{-1}$ was measured at 36 mA $g^{-1}$ current density. In conclusion, the substituted elements slightly enlarged the interlayer spacing favoring a high degree of ordering and improving the ionic transport kinetics, i.e., $Li^+$ ions are faster in doped materials than in the undoped lattice. Less $Li^+/Ni^{2+}$ cation mixing that favors kinetics was also reported for doped NMC materials [26,73]. The poor electrochemical performance of NMC333 is in agreement with a previous report [74], and is due to the important local distortions, which oppose the Li motion and decrease the lattice stability.

## 5. Conclusions

A simple sol-gel method assisted by EDTA as chelator was carried out to prepare $LiNi_{1/3}Mn_{1/3}Co_{1/3}O_2$ cathode materials doped with Al, Mg, Fe, and Zn. The regular morphology and almost identical particle sizes allow for an accurate comparison of electrochemical performances. Considering the intensity ratios of XRD reflections and performing Rietveld refinements, the structural properties showed that not only the degree of cation mixing but also the strain field were sensitive to the doping element. Experimental results revealed that both Mg- and Al-doped NMC333 electrodes delivered the best long-term cyclability and rate capability due to the minimum occupancy of foreign ions in the Li plane. However, Mg doping was the best, because it minimized the local distortions in the lattice, so that the cycle ability was better, even though the discharge capacity in the first cycles was larger with Al-doping.

**Author Contributions:** Conceptualization, A.M.H. and S.I.; methodology, S.I.; investigation, A.E.A.-G. and M.S.; writing—review and editing, C.M.J. and A.M.; supervision, H.E.

**Funding:** This research received no external funding.

**Acknowledgments:** This work contributes to the research performed at CELEST (Center for Electrochemical Energy Storage Ulm-Karlsruhe).

**Conflicts of Interest:** The authors declare no conflict of interest.

## References

1. Julien, C.M.; Mauger, A.; Vijh, A.; Zaghib, K. *Lithium Batteries: Science and Technology*; Springer: Cham, Switzerland, 2016.
2. Goodenough, J.B. Energy storage materials: A perspective. *Energy Storage Mater.* **2015**, *1*, 158–161. [CrossRef]
3. Konarov, A.; Myung, S.-T.; Sun, Y.-K. Cathode materials for future electric vehicles and energy storage systems. *Acs Energy Lett.* **2017**, *2*, 703–708. [CrossRef]
4. Mizushima, K.; Jones, P.C.; Wiseman, P.J.; Goodenough, J.B. $Li_xCoO_2$ ($0 < x < -1$): A new cathode material for batteries of high energy density. *Mater. Res. Bull.* **1980**, *15*, 783–789.
5. Daniel, C.; Mohanty, D.; Li, J.; Wood, D.L. Cathode materials review. *AIP Conf. Proc.* **2014**, *1597*, 26.
6. Senyshyn, A.; Mühlbauer, M.J.; Nikolowski, K.; Pirling, T.; Ehrenberg, H. In-operando neutron scattering studies on Li-ion batteries. *J. Power Sources* **2012**, *203*, 126–129. [CrossRef]
7. Laubach, S.; Laubach, S.; Schmidt, P.C.; Ensling, D.; Schmid, S.; Jaegermann, W.; Thißen, A.; Nikolowski, K.; Ehrenberg, H. Changes in the crystal and electronic structure of $LiCoO_2$ and $LiNiO_2$ upon Li intercalation and de-intercalation. *Phys. Chem. Chem. Phys.* **2009**, *11*, 3278–3289. [CrossRef]
8. Ohzuku, T.; Makimura, Y. Layered lithium insertion material of $Li_{1/2}Mn_{1/2}O_2$: A possible alternative to $LiCoO_2$ for advanced lithium-ion batteries. *Chem. Lett.* **2001**, *30*, 744–745. [CrossRef]
9. Zhecheva, E.; Stoyanova, R. Stabilization of the layered crystal structure of $LiNiO_2$ by Co-substitution. *Solid State Ion.* **1993**, *66*, 143–149. [CrossRef]
10. Julien, C.; El-Farh, L.; Rangan, S.; Massot, M. Studies of $LiNi_{1-y}Co_yO_2$ cathode materials prepared by a citric acid-assisted sol-gel method for lithium batteries. *J. Sol-Gel Sci. Technol.* **1999**, *15*, 63–72. [CrossRef]

11. Chen, Y.; Wang, G.X.; Konstantinov, K.; Liu, H.K.; Dou, S.X. Synthesis and characterization of $LiCo_xMn_yNi_{1-x-y}O_2$ as a cathode material for secondary lithium batteries. *J. Power Sources* **2003**, *119–121*, 184–188. [CrossRef]
12. Ben-Kamel, K.; Amdouni, N.; Abdel-Ghany, A.; Zaghib, K.; Mauger, A.; Gendron, F.; Julien, C.M. Local structure and electrochemistry of $LiNi_yMn_yCo_{1-2y}O_2$ electrode materials for Li-ion batteries. *Ionics* **2008**, *14*, 89–97. [CrossRef]
13. Luo, G.; Zhao, J.; Ke, X.; Zhang, P.; Sun, H.; Wang, B. Structure, electrode voltage and activation energy of $LiMn_xCo_yNi_{1-x-y}O_2$ solid solutions as cathode materials for Li batteries from first-principles. *J. Electrochem. Soc.* **2012**, *159*, A1203–A1208. [CrossRef]
14. MacNeil, D.D.; Lu, Z.; Dahn, J.R. Structure and electrochemistry of $Li[Ni_xCo_{1-2x}Mn_x]O_2$ $(0 \le x \le \frac{1}{2})$. *J. Electrochem. Soc.* **2002**, *149*, A1332–A1336. [CrossRef]
15. Noh, H.-J.; Youn, S.; Yoon, C.S.; Sun, Y.-K. Comparison of the structural and electrochemical properties of layered $Li[Ni_xCo_yMn_z]O_2$ (x = 1/3, 0.5, 0.6, 0.7, 0.8 and 0.85) cathode material for lithium-ion batteries. *J. Power Sources* **2013**, *233*, 121–130. [CrossRef]
16. Jouanneau, S.; Eberman, K.W.; Krause, L.J.; Dahn, J.R. Synthesis, characterization, and electrochemical behavior of improved $Li[Ni_xCo_{1-2x}Mn_x]O_2$ $(0.1 \le x \le 0.5)$. *J. Electrochem. Soc.* **2003**, *150*, A1637–A1642. [CrossRef]
17. Zhang, X.; Jiang, W.J.; Mauger, A.; Gendron, F.; Julien, C.M.; Qilu, R. Minimization of the cation mixing in $Li_{1+x}(NMC)_{1-x}Co_{1/3}O_2$ as cathode material. *J. Power Sources* **2010**, *195*, 1292–1301. [CrossRef]
18. Ohzuku, T.; Makimura, Y. Layered lithium insertion material of $LiNi_{1/3}Co_{1/3}Mn_{1/3}O_2$ for lithium-ion batteries. *Chem. Lett.* **2001**, *30*, 642–643. [CrossRef]
19. Shaju, K.M.; Subba-Rao, G.V.; Chowdari, B.V.R. Performance of layered $Li(Ni_{1/3}Co_{1/3}Mn_{1/3})O_2$ as cathode for Li-ion batteries. *Electrochim. Acta* **2002**, *48*, 145–151. [CrossRef]
20. Hashem, A.M.A.; Abdel-Ghany, A.E.; Eid, A.E.; Trottier, J.; Zaghib, K.; Mauger, A.; Julien, C.M. Study of the surface modification of $LiNi_{1/3}Co_{1/3}Mn_{1/3}O_2$ cathode material for lithium ion battery. *J. Power Sources* **2011**, *196*, 8632–8637. [CrossRef]
21. Kang, K.; Ceder, G. Factors that affect Li mobility in layered lithium transition metal oxides. *Phys. Rev. B* **2006**, *74*, 094105. [CrossRef]
22. Fergus, J.W. Recent developments in cathode materials for lithium ion batteries. *J. Power Sources* **2010**, *195*, 939–954. [CrossRef]
23. Eiler-Rethwish, M.; Winter, M.; Schappacher, F.M. Synthesis, electrochemical investigation and structural analysis of doped $Li[Ni_{0.6}Mn_{0.2}Co_{0.2\text{-}x}M_x]O_2$ ($x$ = 0, 0.05; $M$ = Al, Fe, Sn) cathode materials. *J. Power Sources* **2018**, *387*, 101–107. [CrossRef]
24. Cho, W.; Song, J.H.; Lee, K.-W.; Lee, M.-W.; Kim, H.; Yu, J.-S.; Kim, Y.-J.; Kim, K.J. Improved particle hardness of Ti-doped $LiNi_{1/3}Co_{1/3}Mn_{1/3-x}Ti_xO_2$ as high voltage cathode material for lithium-ion batteries. *J. Phys. Chem. Solids* **2018**, *123*, 271–278. [CrossRef]
25. Li, G.Y.; Huang, Z.L.; Zuo, Z.C.; Zhang, Z.J.; Zhou, H.H. Understanding the trace Ti surface doping on promoting the low temperature performance of $LiNi_{1/3}Co_{1/3}Mn_{1/3}O_2$ cathode. *J. Power Sources* **2015**, *281*, 69–76. [CrossRef]
26. Chen, Q.Y.; Du, C.Q.; Zhang, X.; Tang, Z. Syntheses and characterization of Zr-doped $LiNi_{0.4}Co_{0.2}Mn_{0.4}O_2$ cathode materials for lithium ion batteries. *RSC Adv.* **2015**, *5*, 75248–75253. [CrossRef]
27. Wu, J.; Liu, H.; Ye, X.; Xia, J.; Lu, Y.; Lin, C.; Yu, X. Effect of Nb doping on electrochemical properties of $LiNi_{1/3}Co_{1/3}Mn_{1/3}O_2$ at high cutoff voltage for lithium-ion battery. *J. Alloy. Compd.* **2015**, *644*, 223–227. [CrossRef]
28. Wilcox, J.D.; Patoux, S.; Doeff, M.M. Structure and electrochemistry of $LiNi_{1/3}Co_{1/3-y}M_yMn_{1/3}O_2$ (M = Ti, Al, Fe) positive electrode materials. *J. Electrochem. Soc.* **2009**, *156*, A192–A198. [CrossRef]
29. Liu, L.; Sun, K.; Zhang, N.; Yang, T. Improvement of high-voltage cycling behavior of $Li(Ni_{1/3}Co_{1/3}Mn_{1/3})O_2$ cathodes by Mg, Cr, and Al substitution. *J. Solid State Electrochem.* **2009**, *13*, 1381–1386. [CrossRef]
30. Yang, L.; Ren, F.; Feng, Q.; Xu, G.; Li, X.; Li, Y.; Zhao, E.; Ma, J.; Fan, S. Effect of Cu doping on the structural and electrochemical performance of $LiNi_{1/3}Co_{1/3}Mn_{1/3}O_2$ cathode materials. *J. Electron. Mater.* **2018**, *47*, 3996–4002. [CrossRef]
31. Ding, Y.H.; Ping, Z.; Yong, J. Synthesis and electrochemical properties of $Li[Ni_{1/3}Co_{1/3}Mn_{1/3}]_{1-x}La_xO_2$ cathode materials. *J. Rare Earths* **2007**, *25*, 268–270.
32. Ding, Y.; Zhang, P.; Jiang, Y.; Gao, D. Effect of rare earth elements doping on structure and electrochemical properties of $LiNi_{1/3}Co_{1/3}Mn_{1/3}O_2$ for lithium-ion battery. *Solid State Ion.* **2007**, *178*, 967–971. [CrossRef]

33. Zhang, Y.; Xia, S.; Zhang, Y.; Dong, P.; Yan, X.; Yang, R. Ce-doped $LiNi_{1/3}Co_{(1/3-x/3)}Mn_{1/3}Ce_{x/3}O_2$ cathode materials for use in lithium ion batteries. *Chin. Sci. Bull.* **2012**, *57*, 4181–4187. [CrossRef]
34. Wang, M.; Chen, Y.; Wu, F.; Su, Y.; Chen, L.; Wang, D. Characterization of yttrium substituted $LiNi_{0.33}Mn_{0.33}Co_{0.33}O_2$ cathode material for lithium secondary cells. *Electrochim. Acta* **2010**, *55*, 8815–8820. [CrossRef]
35. Du, H.; Zheng, Y.; Dou, Z.; Zhan, H. Zn-doped $LiNi_{1/3}Co_{1/3}Mn_{1/3}O_2$ composite as cathode material for lithium ion battery: Preparation, characterization, and electrochemical properties. *J. Nanomater.* **2015**, *16*, 339. [CrossRef]
36. Liu, D.; Wang, Z.; Chen, L. Comparison of structure and electrochemistry of Al- and Fe-doped $LiNi_{1/3}Co_{1/3}Mn_{1/3}O_2$. *Electrochim. Acta* **2006**, *51*, 4199–4203. [CrossRef]
37. Wu, F.; Wang, M.; Su, Y.; Bao, L.; Chen, S. A novel layered material of $LiNi_{0.32}Mn_{0.33}Co_{0.33}Al_{0.01}O_2$ for advanced lithium-ion batteries. *J. Power Sources* **2010**, *195*, 2900–2904. [CrossRef]
38. Ren, H.; Li, X.; Peng, Z. Electrochemical properties of $Li[Ni_{1/3}Mn_{1/3}Al_{1/3-x}Co_x]O_2$ as a cathode material for lithium ion battery. *Electrochim. Acta* **2011**, *56*, 7088–7091. [CrossRef]
39. Milewska, A.; Molenda, M.; Molenda, J. Structural, transport and electrochemical properties of $LiNi_{1-y}Co_yMn_{0.1}O_2$ and Al, Mg and Cu-substituted $LiNi_{0.65}Co_{0.25}Mn_{0.1}O_2$ oxides. *Solid State Ion.* **2011**, *192*, 313–320. [CrossRef]
40. Meng, Y.S.; Wu, Y.W.; Hwang, B.J.; Li, Y.; Ceder, G. Combining ab initio computation with experiments for designing new electrode materials for advanced lithium batteries: $LiNi_{1/3}Fe_{1/6}Co_{1/6}Mn_{1/3}O_2$. *J. Electrochem. Soc.* **2004**, *151*, A1134–A1140. [CrossRef]
41. Liu, Q.; Du, K.; Guo, H.; Peng, Z.-D.; Cao, Y.-B.; Hu, G.-R. Structural and electrochemical properties of Co-Mn-Mg multi-doped nickel based cathode materials $LiNi_{0.9}Co_{0.1-x}[Mn_{1/2}Mg_{1/2}]_xO_2$ for secondary lithium ion batteries. *Electrochim. Acta* **2013**, *90*, 350–357. [CrossRef]
42. Lv, C.; Yang, J.; Peng, Y.; Duan, X.; Ma, J.; Li, Q.; Wang, T. 1D Nb-doped $LiNi_{1/3}Co_{1/3}Mn_{1/3}O_2$ nanostructures as excellent cathodes for Li-ion battery. *Electrochim. Acta* **2019**, *297*, 258–266. [CrossRef]
43. Lin, Y.-K.; Lu, C.-H. Preparation and electrochemical properties of layer structured $LiNi_{1/3}Co_{1/3}Mn_{1/3-y}Al_yO_2$. *J. Power Sources* **2009**, *189*, 353–358. [CrossRef]
44. Zhou, F.; Zhao, X.; Dahn, J.R. Synthesis, electrochemical properties, and thermal stability of Al-doped $LiNi_{1/3}Mn_{1/3-y}Co_{(1/3-z)}Al_zO_2$ positive electrode materials. *J. Electrochem. Soc.* **2009**, *156*, A343–A347. [CrossRef]
45. Zhou, F.; Zhao, X.; Jiang, J.; Dahn, J.R. Advantages of simultaneous substitution of Co in $Li[Ni_{1/3}Mn_{1/3}Co_{1/3}]O_2$ by Ni and Al. *Electrochem. Solid-State Lett.* **2009**, *12*, A81–A83. [CrossRef]
46. Hashem, A.M.; Abdel-Ghany, A.E.; Abuzeid, H.M.; El-Tawil, R.S.; Indris, S.; Ehrenberg, H.; Mauger, A.; Julien, C.M. EDTA as chelating agent for sol-gel synthesis of spinel $LiMn_2O_4$ cathode material for lithium batteries. *J. Alloy. Compd.* **2018**, *737*, 758–766. [CrossRef]
47. Hashem, A.M.; El-Taweel, R.S.; Abuzeid, H.M.; Abdel-Ghany, A.E.; Eid, A.E.; Groult, H.; Mauger, A.; Julien, C.M. Structural and electrochemical properties of $LiNi_{1/3}Co_{1/3}Mn_{1/3}O_2$ material prepared by a two-step synthesis via oxalate precursor. *Ionics* **2012**, *18*, 1–9. [CrossRef]
48. El-Mofid, W.; Ivanov, S.; Konkin, A.; Bund, A. A high-performance layered transition metal oxide cathode material obtained by simultaneous aluminum and iron cationic substitution. *J. Power Sources* **2014**, *268*, 122–141. [CrossRef]
49. Ohzuku, T.; Ueda, A.; Nagayama, M. Electrochemistry and structural chemistry of $LiNiO_2$ ($R\bar{3}m$) for 4 volt secondary lithium cells. *J. Electrochem. Soc.* **1993**, *140*, 1862–1870. [CrossRef]
50. Dahn, J.R.; von Sacken, U.; Michal, C.A. Structure and electrochemistry of $Li_{1\pm y}NiO_2$ and a new $Li_2NiO_2$ phase with the $Ni(OH)_2$ structure. *Solid State Ion.* **1990**, *44*, 87–97. [CrossRef]
51. Reimers, J.N.; Rossen, E.; Jones, C.D.; Dahn, J.R. Structure and electrochemistry of $Li_xFe_yNi_{1-y}O_2$. *Solid State Ion.* **1993**, *61*, 335–344. [CrossRef]
52. Chang, C.-C.; Kim, J.Y.; Kumta, P.N. Divalent cation incorporated $Li_{(1+x)}MMg_xO_{2(1+x)}$ ($M = Ni_{0.75}Co_{0.25}$): Viable cathode materials for rechargeable lithium-ion batteries. *J. Power Sources* **2000**, *89*, 56–63. [CrossRef]
53. Kim, G.-H.; Myung, S.-T.; Kim, H.-S.; Sun, Y.-K. Synthesis of spherical $Li[Ni_{(1/3-z)}Co_{(1/3-z)}Mn_{(1/3-z)}Mg_z]O_2$ as positive electrode material for lithium-ion battery. *Electrochim. Acta* **2006**, *51*, 2447–2453. [CrossRef]
54. Shannon, R. Revised effective ionic radii and systematic studies of interatomic distances in halides and chacogenides. *Acta Cryst. A* **1976**, *32*, 751–767. [CrossRef]
55. Tran, N.; Croguennec, L.; Jordy, C.; Biensan, P.; Delmas, C. Influence of the synthesis route on the electrochemical properties of $LiNi_{0.425}Mn_{0.425}Co_{0.15}O_2$. *Solid State Ion.* **2005**, *176*, 1539–1547. [CrossRef]

56. Zhang, X.; Mauger, A.; Lu, Q.; Groult, H.; Perrigaud, L.; Gendron, F.; Julien, C.M. Synthesis and characterization of $LiNi_{1/3}Mn_{1/3}Co_{1/3}O_2$ by wet chemical method. *Electrochim. Acta* **2010**, *55*, 6440–6449. [CrossRef]
57. Grey, C.P.; Dupre, N. NMR studies of cathode materials for lithium-ion rechargeable batteries. *Chem. Rev.* **2004**, *104*, 4493–4512. [CrossRef] [PubMed]
58. Cahill, L.S.; Yin, S.C.; Samoson, A.; Heinmaa, I.; Nazar, L.F.; Goward, G.R. $^6$Li NMR studies of cation disorder and transition metal ordering in $Li[Ni_{1/3}Mn_{1/3}Co_{1/3}]O_2$ using ultrafast magic angle spinning. *Chem. Mater.* **2005**, *17*, 6560–6566. [CrossRef]
59. Han, B.; Key, B.; Lapidus, S.H.; Garcia, J.C.; Iddir, H.; Vaughey, J.T.; Dogan, F. From coating to dopant: How the transition metal composition affects alumina coatings on Ni-rich cathodes. *Acs Appl. Mater. Interfaces* **2017**, *9*, 41291–41302. [CrossRef]
60. Ye, S.; Xia, Y.; Zhang, P.; Qiao, Z. Al, B, and F doped $LiNi_{1/3}Co_{1/3}Mn_{1/3}O_2$ as cathode material of lithium-ion batteries. *J. Solid State Electrochem.* **2007**, *11*, 805–810. [CrossRef]
61. Riley, L.A.; Van Atta, S.; Cavanagh, A.S.; Yan, Y.; George, S.M.; Liu, P.; Dillon, A.C.; Lee, S.-H. Electrochemical effects of ALD surface modification on combustion synthesized $LiNi_{1/3}Mn_{1/3}Co_{1/3}O_2$ as a layered-cathode material. *J. Power Sources* **2011**, *196*, 3317–3324. [CrossRef]
62. Yabuuchi, N.; Makimura, Y.; Ohzuku, T. Solid-state chemistry and electrochemistry of $LiCo_{1/3}Ni_{1/3}Mn_{1/3}O_2$ for advanced lithium-ion batteries: III. Rechargeable capacity and cycleability. *J. Electrochem. Soc.* **2007**, *154*, A314–A321. [CrossRef]
63. Li, Z.-Y.; Zhang, H.-L. The improvement for the electrochemical performances of $LiNi_{1/3}Mn_{1/3}Co_{1/3}O_2$ cathode materials for lithium-ion batteries by both the Al-doping and an advanced synthetic method. *Int. J. Electrochem. Sci.* **2019**, *14*, 3524–3534. [CrossRef]
64. Kobayashi, T.; Kawasaki, N.; Kobayashi, Y.; Shono, K.; Mita, Y.; Miyashiro, H. A method of separating the capacities of layer and spinel compounds in blended cathode. *J. Power Sources* **2014**, *245*, 1–6. [CrossRef]
65. Luo, W.; Zhou, F.; Zhao, X.; Lu, Z.; Li, X.; Dahn, J.R. Synthesis, Characterization, and Thermal Stability of $LiNi_{1/3}Mn_{1/3}Co_{1/3-z}Mg_zO_2$, $LiNi_{1/3-z}Mn_{1/3}Co_{1/3}Mg_zO_2$, and $LiNi_{1/3}Mn_{1/3-z}Co_{1/3}Mg_zO_2$. *Chem. Mater.* **2010**, *22*, 1164–1172. [CrossRef]
66. Zhang, Q.; Zheng, Y.; Zhong, S. Preparation and electrochemical performance of Mg-doped $LiNi_{0.4}Co_{0.2}Mn_{0.4}O_2$ cathode materials by urea co-precipitation. *Key Eng. Mater.* **2012**, *519*, 152–155. [CrossRef]
67. Chang, C.C.; Kim, J.Y.; Kumta, P.N. Synthesis and electrochemical characterization of divalent cation-incorporated lithium nickel oxide. *J. Electrochem. Soc.* **2000**, *147*, 1722–1729. [CrossRef]
68. D'Epifanio, A.; Croce, F.; Ronci, F.; Rossi-Albertini, V.; Traversa, E.; Scrosati, B. Effect of $Mg^{2+}$ doping on the structural, thermal, and electrochemical properties of $LiNi_{0.8}Co_{0.16}Mg_{0.04}O_2$. *Chem. Mater.* **2004**, *16*, 3559–3564. [CrossRef]
69. Martha, S.K.; Hadar Sclar, K.; Framowitz, Z.S.; Kovacheva, D.; Saliyski, N.; Gofer, Y.; Sharona, P.; Golikc, E.; Markovskya, B.; Aurbach, D. A comparative study of electrodes comprising nanometric and submicron particles of $LiNi_{0.50}Mn_{0.50}O_2$, $LiNi_{0.33}Mn_{0.33}Co_{0.33}O_2$, and $LiNi_{0.40}Mn_{0.40}Co_{0.20}O_2$ layered compounds. *J. Power Sources* **2009**, *189*, 248–255. [CrossRef]
70. Ceder, G.; Chiang, Y.M.; Sadoway, D.R.; Aydinol, M.K.; Jang, Y.I.; Huang, B. Identification of cathode materials for lithium batteries guided by first-principles calculations. *Nature* **1998**, *392*, 694–696. [CrossRef]
71. Castro-Couceiro, A.; Castro-Garcia, S.; Senaris-Rodriguez, M.A.; Soulette, F.; Julien, C. Influence of aluminium doping on the properties $LiCoO_2$ and $LiNi_{0.5}Co_{0.5}O_2$ oxides. *Solid State Ion.* **2003**, *156*, 15–26. [CrossRef]
72. Samarasingha, P.B.; Wijayasinghe, A.; Behm, M.; Dissanayake, L.; Lindbergh, G. Development of cathode materials for lithium ion rechargeable batteries based on the system $Li(Ni_{1/3}Mn_{1/3}Co_{(1/3-x)}M_x)O_2$, (M = Mg, Fe, Al and x = 0.00 to 0.33). *Solid State Ion.* **2014**, *268*, 226–230. [CrossRef]
73. Wang, M.; Chen, Y.; Wu, F.; Su, Y.F.; Chen, L. Improvement of the electrochemical performance of $LiNi_{0.33}Mn_{0.33}Co_{0.33}O_2$ cathode material by chromium doping. *Sci. China Technol. Sci.* **2010**, *53*, 3214–3220. [CrossRef]
74. Prado, G.; Fournès, L.; Delmas, C. On the $Li_xNi_{0.70}Fe_{0.15}Co_{0.15}O_2$ system: An X-ray diffraction and Mössbauer study. *J. Solid State Chem.* **2001**, *159*, 103–112. [CrossRef]

*Article*

# Synthesis and Characterization of $LiFePO_4$–PANI Hybrid Material as Cathode for Lithium-Ion Batteries

**Cesario Ajpi [1,2,*], Naviana Leiva [1], Max Vargas [1], Anders Lundblad [3], Göran Lindbergh [2] and Saul Cabrera [1,*]**

1 Department of Inorganic Chemistry and Materials Science/Advanced Materials, IIQ Chemical Research Institute, UMSA Universidad Mayor de San Andres, La Paz 303, Bolivia; nindeleivqmc@gmail.com (N.L.); vargmax@gmail.com (M.V.)

2 Department of Chemical Engineering, Applied Electrochemistry, KTH Royal Institute of Technology, SE-10044 Stockholm, Sweden; gnli@kth.se

3 Division of Safety and Transport/Electronics, RISE, Research Institutes of, Sweden, SE-50462 Borås, Sweden; anders.lundblad@ri.se

* Correspondence: cesario.ajpi@gmail.com (C.A.); saulcabreram@hotmail.com (S.C.)

Received: 28 April 2020; Accepted: 15 June 2020; Published: 24 June 2020

**Abstract:** This work focuses on the synthesis of $LiFePO_4$–PANI hybrid materials and studies their electrochemical properties (capacity, cyclability and rate capability) for use in lithium ion batteries. PANI synthesis and optimization was carried out by chemical oxidation (self-assembly process), using ammonium persulfate (APS) and $H_3PO_4$, obtaining a material with a high degree of crystallinity. For the synthesis of the $LiFePO_4$–PANI hybrid, a thermal treatment of $LiFePO_4$ particles was carried out in a furnace with polyaniline (PANI) and lithium acetate (AcOLi)-coated particles, using $Ar/H_2$ atmosphere. The pristine and synthetized powders were characterized by XRD, SEM, IR and TGA. The electrochemical characterizations were carried out by using CV, EIS and galvanostatic methods, obtaining a capacity of 95 $mAhg^{-1}$ for PANI, 120 $mAhg^{-1}$ for $LiFePO_4$ and 145 $mAhg^{-1}$ for $LiFePO_4$–PANI, at a charge/discharge rate of 0.1 C. At a charge/discharge rate of 2 C, the capacities were 70 $mAhg^{-1}$ for $LiFePO_4$ and 100 $mAhg^{-1}$ for $LiFePO_4$–PANI, showing that the PANI also had a favorable effect on the rate capability.

**Keywords:** PANI, $LiFePO_4$; conducting polymers; hybrid materials; lithium-ion batteries

## 1. Introduction

Since Sony developed and commercialized the first lithium-ion batteries based on $LiCoO_2$ in 1990, new materials and different applications have driven research and development. One direction has been in the development of materials with improved electrochemical properties (specific energy, energy density, rate capability and cyclability) for hybrid and electric vehicles and energy storage systems. Materials that have been extensively studied include inorganic materials such as $LiFePO_4$ [1,2] and $LiMn_2O_4$ [3,4]; organic materials such as $Li_4C_6O_6$, quinones and anthraquinones [5]; polyaniline (PANI); and others.

$LiFePO_4$ has two main disadvantages: low electronic conductivity and slow Li-ion diffusion due to its 1D channel for Li extraction. Efforts have been made to improve the electrochemical performance by carbon coating [6], cation doping [7,8] or minimizing the particle size [9,10]. Carbon coating is a common method to enhance the electronic conductivity of $LiFePO_4$. The carbon is electrochemically inactive, but its incorporation in a $LiFePO_4$ electrode can influence the capacity and cyclability. Substituting the inactive carbon black and binder typically present in composite electrodes with an electrochemically active polymer like polypyrrole (PPy) or polyaniline (PANI) has been proposed [11–17], to enhance the electrochemical performance. PANI is a promising conducting polymer due to its facile synthesis, environmental stability and tunable physical and electrochemical properties controlled by oxidation and protonation [18]. PANI is electrochemically active in the range of 2.0–3.8 V, which overlaps with the redox couple of $LiFePO_4$. Hybrid electrodes for lithium-ion batteries incorporating PANI with $LiMn_2O_4$ [19], $MnO_2$ [20], $V_2O_5$ [21] and $Li(Mn_{1/3}Ni_{1/3}Fe_{1/3})O_2$ [22] have been reported. PANI shows compatible polarity with the electrolyte via the formation of H bonds. PANI can mediate the polarity difference between the cathode particles and the electrolyte, promoting electrolyte permeation into the surface of the active particles and hence enhancing Li-ion extraction/insertion during a charge/discharge process. Incorporating $LiFePO_4$ with conductive PANI is thereby an attractive route to improve both electronic conductivity and Li-ion diffusion. The capacity of PANI is usually higher than that of PPy or PT, but depends on doping [23]. Various dopants have been used to improve the physical and chemical properties of PANI. Among them, salts such as $LiClO_4$, $LiBF_4$, $LiPF_6$ and $Zn(ClO_4)_2$ have received much attention, and their application in rechargeable lithium-ion batteries has been extensively studied [24].

In general, PANI and $LiFePO_4$–PANI are synthesized via oxidative polymerization in acidic media. For the synthesis of PANI, various oxidants, such as ammonium peroxydisulfate (APS) [25], hydrogen peroxide [26], ferric chloride [18] and ferric sulfate [27], have been used. With APS as the oxidant, PANI can be successfully doped with inorganic acid (e.g., HCl, $H_2SO_4$ and $H_3PO_4$) [28] or organic acid (e.g., dicarboxylic acid) [18]. The most common method for the synthesis of $LiFePO_4$–PANI is the shelf-assembly process, PANI is incorporated with $LiFePO_4$ particles through simultaneous chemical polymerization with APS as the oxidizer, and aniline and an inorganic acid as dopants [12].

The overall chemical and electrochemical properties can be improved with advanced hybrid materials combining inorganic and organic materials. For the design of this type of hybrid material, the interactions between the inorganic and organic parts are key (supramolecular chemistry); this opens up an immense number of possibilities for different combinations [29].

This work is focused on the synthesis, structural characterization and electrochemical characterization of $LiFePO_4$–PANI as a cathode material for lithium-ion batteries. PANI was prepared via a self-assembly process, commercial $LiFePO_4$ was used and a $LiFePO_4$–PANI hybrid material was synthetized by a solid-state reaction. The main challenge in the synthesis of $LiFePO_4$–PANI via self-assembly process is $LiFePO_4$ dissolution by the acid (e.g., HCl, $H_2SO_4$ and $H_3PO_4$) and oxidation of the $LiFePO_4$ to $FePO_4$ by APS, resulting in low reaction yields. The method used for the synthesis of $LiFePO_4$–PANI in this study does not use APS and acids, which is advantageous in terms of reaction yield. PANI and AcOLi were simultaneously incorporated with $LiFePO_4$ particles through a thermal treatment. Additionally, the method incorporates Li-ions in the PANI structure as dopants. XRD, FTIR and TG were used for the characterization of $LiFePO_4$, PANI and $LiFePO_4$–PANI, and the surface morphology was studied by using SEM. The electrochemical behavior and discharge/charge performance were investigated by EIS, CV and charge/discharge processes. Rate capability and cyclability were measured for the $LiFePO_4$–PANI and showed improvement when compared to the pure materials.

## 2. Materials and Methods

### 2.1. Materials Synthesis of PANI

First, 25 mmol of ammonium persulfate (APS) (98%, Sigma-Aldrich, St. Louis, MS, USA) was dissolved in 50 mL of distilled water. Then, 20 mmol of aniline was dissolved in 50 mL of 1M of $H_3PO_4$ (ACS reagent, ≥85 wt%, Sigma-Aldrich, St. Louis, MS, USA) in aqueous media. Both solutions were prepared at room temperature. The APS solution was added to the solution containing the aniline. The mixture was magnetically stirred for 1 h, at 5 °C. The dark blue precipitate of emeraldine base (EB) was collected by filtration and washed 5 times with distilled water and 5 times with ethanol and dried under vacuum, at 60 °C, for 8 h.

### 2.2. Synthesis of $LiFePO_4$–PANI

The synthesis of $LiFePO_4$–PANI with ~25% polyaniline was carried out by using 0.75 g of commercial $LiFePO_4$ (Phostech Lithium Inc., St-Bruno-de-Montarville, Quebec, Canada, coated with 2% of C), 0.25 g of synthetized PANI (i.e., EB) and 0.1675 g of lithium acetate (AcOLi) (99.95%, Sigma-Aldrich, St. Louis, MS, USA). The amounts of the compounds were mixed in a mortar for 0.5 h. After that, the mixture was transferred to a crucible and thermally treated at 300 °C, for 1 h, in an Ar/$H_2$ (90/10) atmosphere. A black powder was obtained with a weight yield of 99.98%.

### 2.3. Structural Characterization

The $LiFePO_4$–PANI was characterized by X-ray powder diffraction (PXRD), using an X'Pert3 PANalytical diffractometer with a scan rate of 0.02 °/s, by Cu–K$\alpha$ radiation ($\lambda$ = 1.5406 Å, 45 kV, 20 mA). X'pert Highscore software (PANalytical. B. V, Lelyweg, the Netherlands) was used for the identification of the phases. The morphologies of the samples were determined by S-4800 series field-emission high-resolution scanning electron microscope (SEM) (Hitachi, Tokyo, Japan), equipped with an EDS detector (Oxford Instruments, Abingdon, UK). The powder samples were mounted in the sample holder and sputtered with a thin layer of Pt/Pd. Fourier transform infrared spectroscopy (FTIR) was done on a Spectrum 100 spectrometer (Perkin Elmer), in the wavelength range of 4000–400 $cm^{-1}$, with a resolution of 1 $cm^{-1}$. Thermogravimetric analysis (TGA) (LABSYS evo STA 1150, SETARAM Instrumentation, Caluire-et-Cuire, France) was carried out on 20 mg of specimen, from room temperature to 800 °C, with a heating rate of 2 °C/min.

### 2.4. Electrode Preparation and Electrochemical Characterization

The active cathode material was made by preparing an 80:10:10 mixture of active material (LFP–PANI), conducting agent (Super P) and binder (PVDF), respectively. The components of the mixture were weighed and subsequently ground for 15 min. Then N-methylpyrrolidone (NMP) was added as solvent, and the mixture was magnetically stirred for 1 h. The electrodes were tape-casted by a doctor blade process on an Al foil with a slit height of 100 µm. The electrochemical testing was done in a bottom-type test cell of 0.9 $cm^2$ active area, using metallic Li as the counter and reference electrode, Celgard (2325) as separator and a solution of 1 M $LiPF_6$ dissolved in EC-DMC (1:1 vol) as electrolyte. The test cells were subjected to charge–discharge processes at a rate of 0.1 C between 2.5 and 4.2 V vs. $Li^+/Li^0$ and electrochemical impedance spectroscopy (EIS) measurement in the frequency range between 100 kHz and 0.1 Hz, with an AC voltage amplitude of 10 mV, using a Gamry potentiostat 600.

## 3. Results and Discussion

### 3.1. Synthesis of PANI and $LiFePO_4$–PANI

The synthesis of PANI was performed by chemical oxidation of aniline with APS; the reaction is shown in Figure 1.

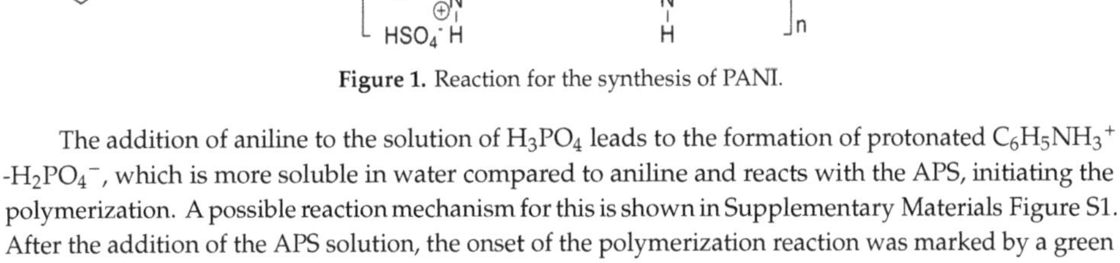

$NH_2$ 4n + 5n $(NH_4)_2S_2O_8$ $\xrightarrow{H^+}$ [H $HSO_4^-$ N⊕ ... N ... H–N ... ⊕N $HSO_4^-$ H ... N H]$_n$ + 3n $H_2SO_4$ + 5n $(NH_4)_2SO_4$

**Figure 1.** Reaction for the synthesis of PANI.

The addition of aniline to the solution of $H_3PO_4$ leads to the formation of protonated $C_6H_5NH_3^+$ -$H_2PO_4^-$, which is more soluble in water compared to aniline and reacts with the APS, initiating the polymerization. A possible reaction mechanism for this is shown in Supplementary Materials Figure S1. After the addition of the APS solution, the onset of the polymerization reaction was marked by a green coloration due to the formation of a chemical species of the polyaniline called ES (emeraldine salt). Later, the color changed to blue due to the formation of the chemical species EB (emeraldine base); this change was controlled by the $H^+$ concentration (pH). The structures of these species are shown in Figure 2.

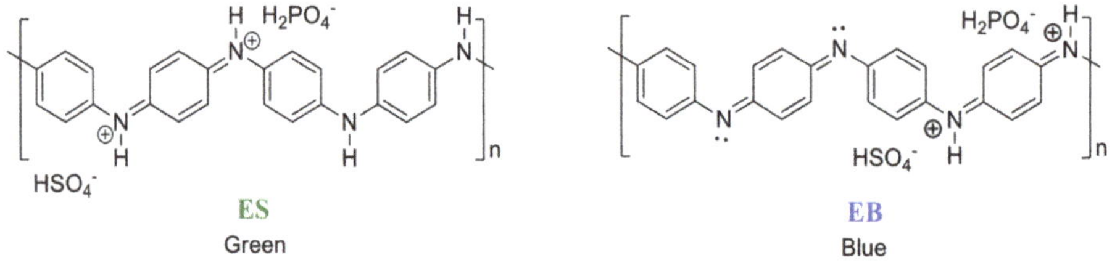

**Figure 2.** Structure of ES (emeraldine salt) and EB (emeraldine base).

The reaction between $LiFePO_4$, PANI and AcOLi starts during mixing, noted by a characteristic smell of acetic acid and a change in the color of the PANI from blue to green. The observed change of color was attributed to the reaction in Figure 3.

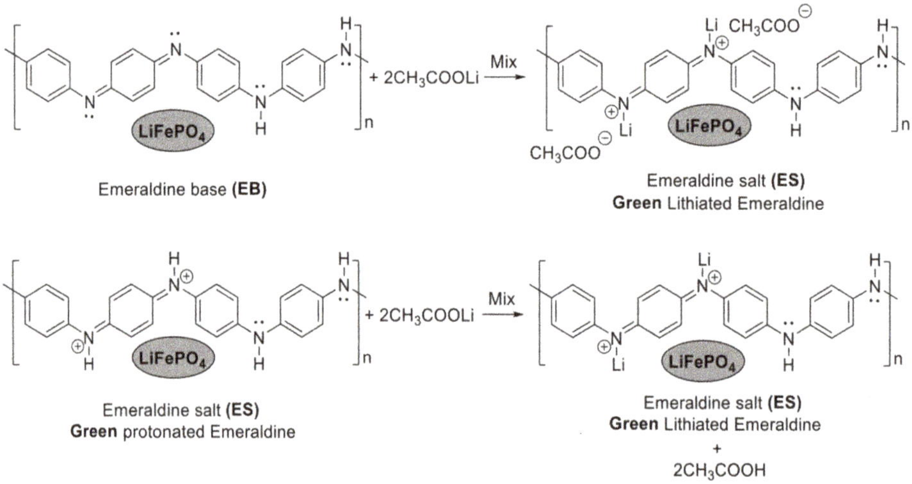

**Figure 3.** Reactions of ES (emeraldine salt) and EB (emeraldine base) with AcOLi.

EB and protonated ES in the PANI structure can be lithiated by reaction with AcOLi, forming (in both cases) lithiated ES. During thermal treatment at 300 °C for 1 h in an inert atmosphere of $Ar/H_2$, the AcOH is removed by evaporation.

The synthesis of the $LiFePO_4$–PANI was carried out at 300 °C. During the thermal treatment, the PANI was converted by a chemical reaction between two chains of PANI [15], as shown in Figure 4.

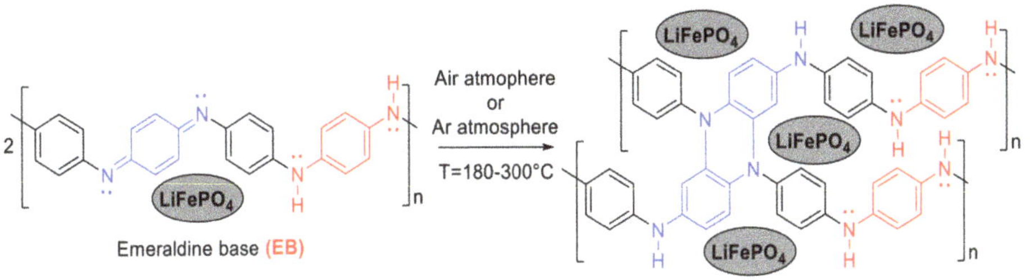

**Figure 4.** Conversion reaction and structure of crosslinked PANI. A quinoid unit of the PANI chain, shown in blue, reacts with a quinoid unit of another PANI chain, to form crosslinked PANI.

## *3.2. PXRD Characterizations of $LiFePO_4$, PANI and $LiFePO_4$–PANI*

The dark blue PANI powder was characterized by PXRD. The diffraction patterns of the product and the starting powders are shown in Figure 5.

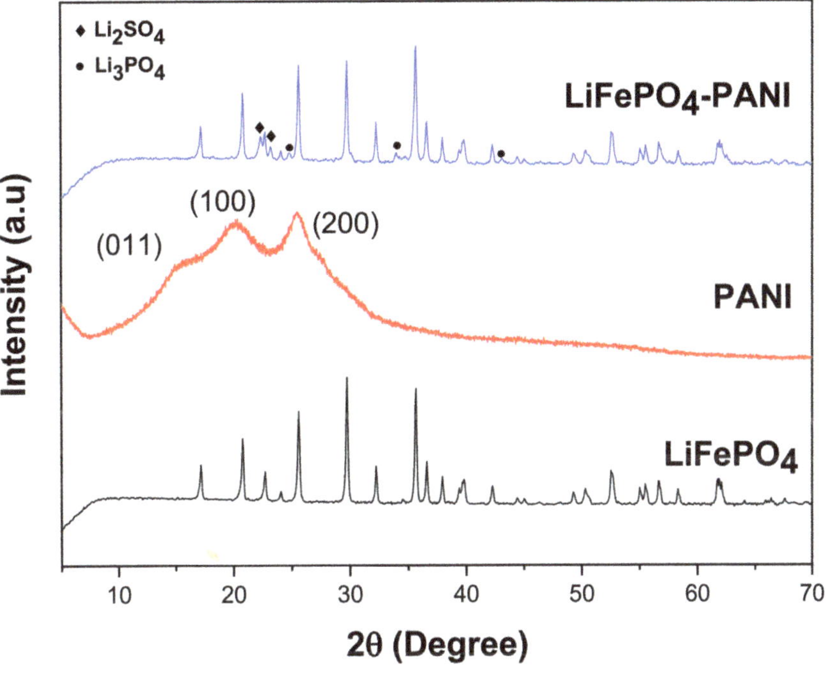


**Figure 5.** Diffraction patterns of the $LiFePO_4$, PANI and $LiFePO_4$–PANI samples.

The diffraction pattern shows a higher crystallinity of the pure PANI than normally reported. In general, the crystallinity of PANI depends of the methods of synthesis and dopants [30,31]. The diffraction planes (011) correspond to the orientation parallel to the polymer chain. The planes (100) correspond to the parallel and perpendicular periodicity of the polymer chain. The planes (200) correspond to the periodicity perpendicular along the chain [11]. The XRD diffraction pattern of the $LiFePO_4$–PANI composite shows diffraction peaks corresponding to group Pnmb for $LiFePO_4$. A low amount of phase impurity is detected as $Li_2SO_4$ (01-075-0929) and $Li_3PO_4$ (01-071-1528). These impurities originate from the reaction between $Li^+$ from the AcOLi and $HSO_4^-$ and $H_2PO_4^-$ in the structure of PANI. A proposed reaction for the formation of $Li_2SO_4$ and $Li_3PO_4$ is shown in Figure 6. The diffraction pattern of PANI in the composite material is not visible, due the lower relative intensity of its diffraction peaks compared to $LiFePO_4$. This decreased intensity is due to the small amount and possibly also lesser crystallinity of the PANI deposited on the particles.

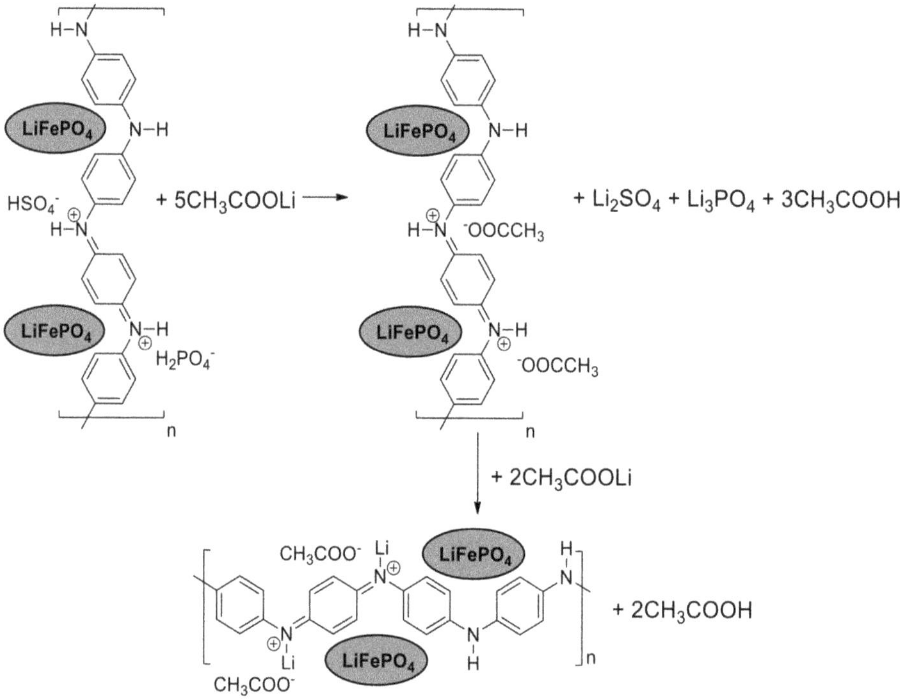

Figure 6. Reactions of the $HSO_4^-$ and $H_2PO_4^-$ groups in the PANI with AcOLi.

### 3.3. Morphologies of $LiFePO_4$, PANI and $LiFePO_4$–PANI

Figure 7a–c shows SEM images of PANI, LFP and LFP–PANI, respectively. The resulting PANI powder prepared by a self-assembly process has globular morphology, with aggregates having an average diameter of 2.75 μm, estimated from Supplementary Materials Figure S2. The primary particles have an average diameter of 310 nm and a globular morphology, which are characteristic of PANI synthesized by chemical oxidation, as shown in Figure 7a and Supplementary Materials Figure S2. In Figure 7b, a distribution of LFP particle size can be seen with an estimated average diameter of 300 nm. The LFP–PANI particles have an estimated average diameter of 180 nm, as seen in Figure 7c. The particles were split via milling and stirring in a mortar before the thermal treatment and were homogeneously coated by PANI. This is an advantage for the diffusion of electrolyte and $Li^+$ into the active particles and leads to a good rate capability of the material [12].

The particle size and shape of the $LiFePO_4$–PANI composite particles are similar to $LiFePO_4$ particles, but the surfaces morphologies are different. Figure 7c shows the expected coating on the composite $LiFePO_4$ particles. Scanning electron microscopy (SEM) imaging coupled with EDS mapping provides a qualitative and semiquantitative elemental analysis of the $LiFePO_4$–PANI powder and is presented in Figure 8.

There is good correlation between the PXRD and the experimental EDS (Figure 9), showing the presence of C, N, P, S and Fe. The maps of N, S and C indicate homogenous distribution of PANI on the $LiFePO_4$ particles. However, some flakes of PANI extending from the particle surfaces can also be observed. The mapping of S shows a low concentration attributed to a small amount of $Li_2SO_4$. The EDS spectrum of the synthesized PANI is shown in Supplementary Materials Figure S3.

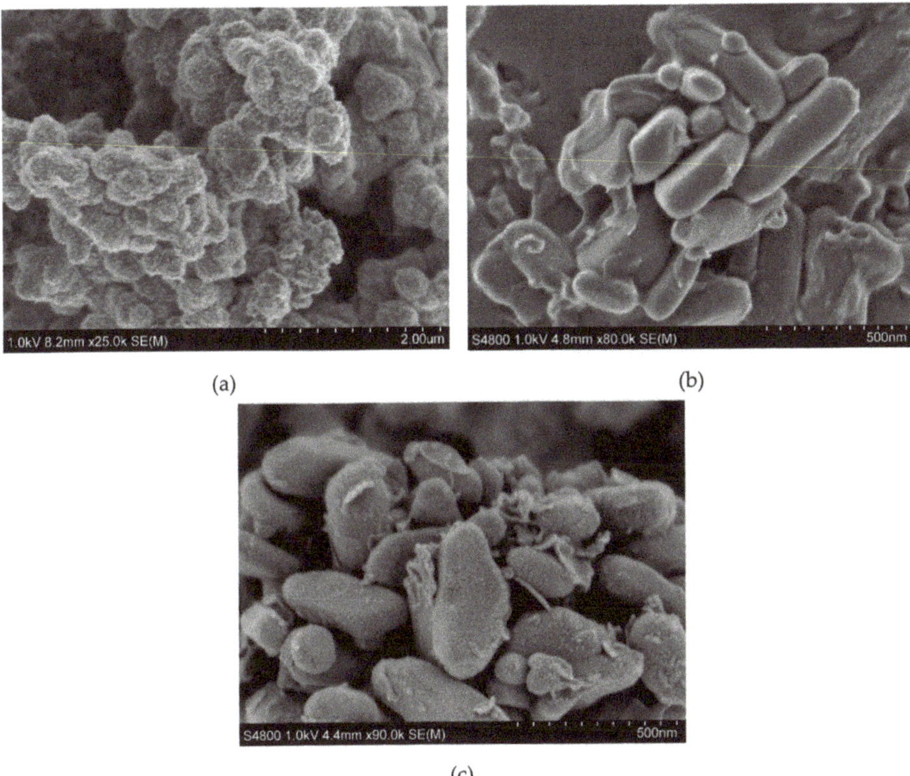


(a) (b)

(c)

**Figure 7.** SEM images for (**a**) synthesized PANI, showing the agglomerates; (**b**) $LiFePO_4$, showing the particles; and (**c**) $LiFePO_4$–PANI hybrid.

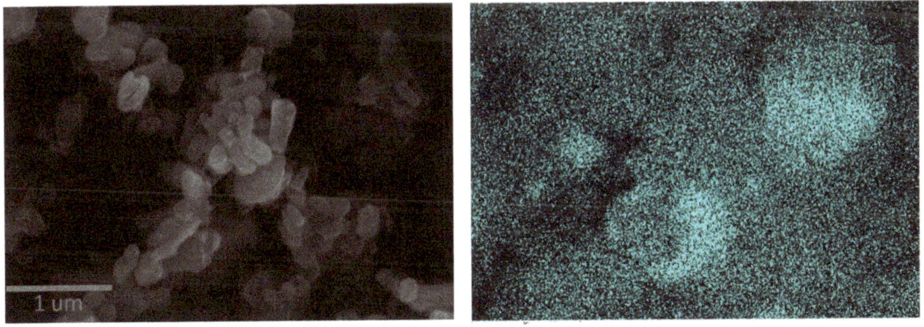


Mapping Element Fe

**Figure 8.** *Cont.*

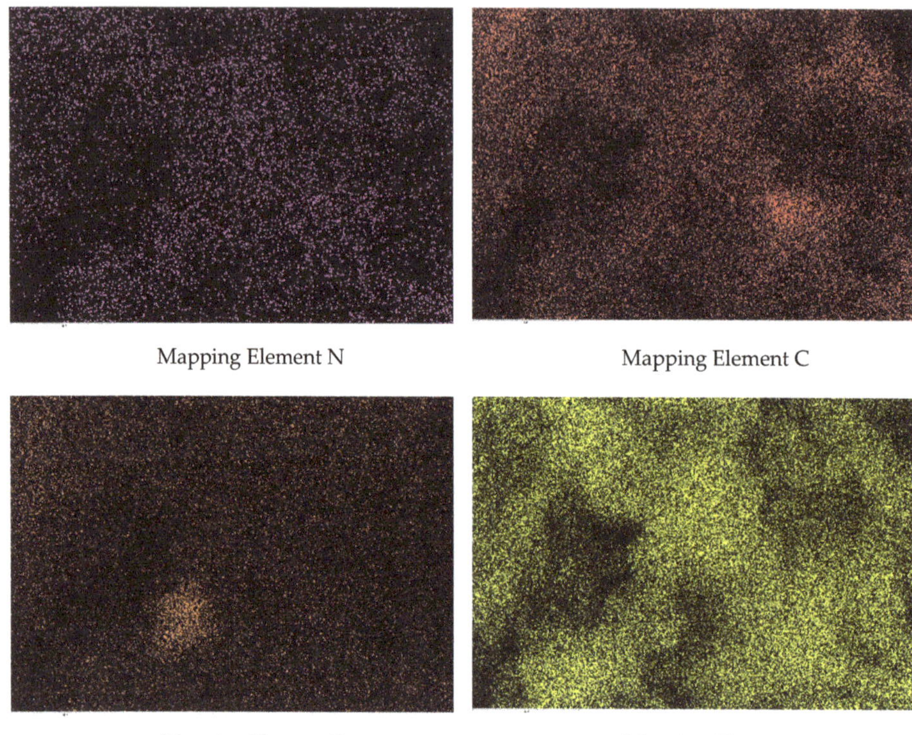


**Figure 8.** Scanning electron microscopy (SEM) of $LiFePO_4$–PANI coupled with energy dispersive spectroscopy (EDS).

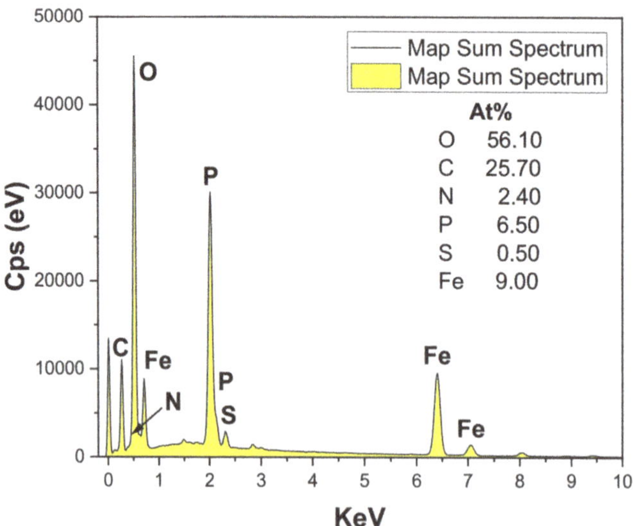


**Figure 9.** Energy dispersive microscopy (EDS) spectrum for $LiFePO_4$–PANI powder.

### 3.4. FT-Infrared Spectroscopy

The IR spectrum of $LiFePO_4$ in Figure 10 shows four fundamental modes; the bands around 1136, 1092, 1058 and 977 $cm^{-1}$ correspond to the P–O antisymmetric stretching vibration of the olivine phosphate groups [32,33]. The bands around 647 and 632 $cm^{-1}$ correspond to the O–P–O symmetric and anti-symmetric stretching vibrations [34]. The IR spectrum of PANI (Supplementary Materials Figure S4) shows the presence of a broad band around 1050 $cm^{-1}$, which is known as an "electronic band" being associated with the doped form of polyaniline, specifically the vibration mode of the -$NH^+$= structures [35].

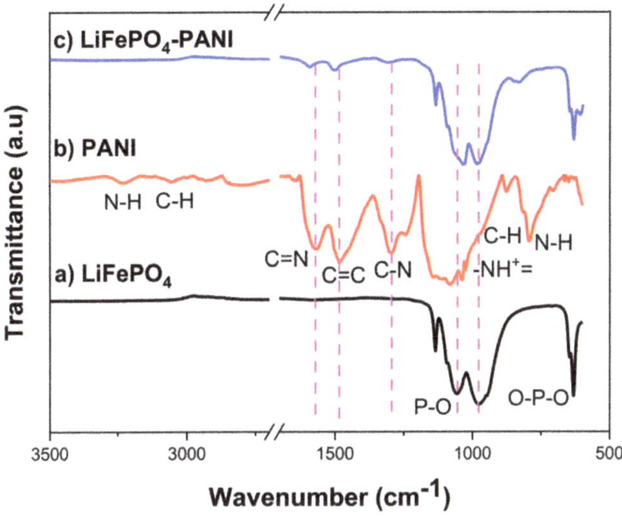


**Figure 10.** FTIR spectra for: (**a**) $LiFePO_4$, (**b**) PANI and (**c**) $LiFePO_4$–PANI.

The spectra also show characteristic peaks at about 1575 $cm^{-1}$ for C = N stretching mode of the quinonoid rings (quinoid unit, Figure 11), 1486 $cm^{-1}$ for C = C stretching mode of benzenoid rings (benzoid unit, Figure 10) and 1299 $cm^{-1}$ for C-N stretching mode. The vibrational bands at about 1144 and 708 $cm^{-1}$ are assigned to the aromatic ring in-plane and out-of-plane C-H bending [36,37]. The peak at 796 $cm^{-1}$ is assigned to N-H wag vibration, characteristic of primary and secondary amines only. The peak at 3229 $cm^{-1}$ is assigned to N-H stretching vibration, which is present in the case of PANI-EB. The structure of PANI is generally represented by the following structure, where X = 0.5 for the polyemeraldine salt:

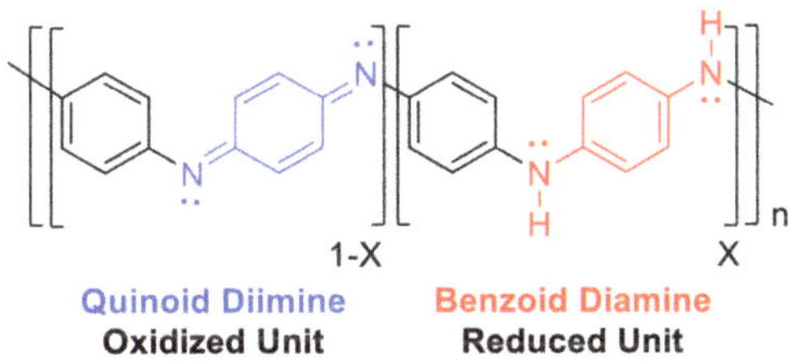

**Figure 11.** Oxidized and reduced units.

The IR spectrum of $LiFePO_4$–PANI shows peak shifts in the broad bands ($LiFePO_4$ band to $LiFePO_4$–PANI band); 1092 to 1194, 1058 to 1031 and 977 to 983 $cm^{-1}$ correspond to the P–O antisymmetric stretching vibration compared to $LiFePO_4$. This is probably due to hydrogen bonding

(P-O—H-N) between $LiFePO_4$ and PANI. The O–P–O symmetric and anti-symmetric stretching vibrations at 632 $cm^{-1}$ and the band at 647 $cm^{1}$ related to P vibration remain unchanged. A new peak appears at 610 $cm^{-1}$, attributed to the vibration of O—H in the interaction O–P–O—H-N.

The vibration mode at 1136 $cm^{-1}$ associated with -$NH^{+}$= structures of polyaniline is overlapped with the P-O vibration. The peak changes compared to polyaniline from 1299 to 1307 $cm^{-1}$ for the C-N stretching, 1575 to 1592 $cm^{-1}$ for the C = N stretching mode of the quinonoid rings and 1486 to 1505 $cm^{-1}$ for the C = C stretching mode of benzenoid rings reveal that the polyaniline in the $LiFePO_4$–PANI has more quinoid units than the PANI. This is probably because the chain of PANI deposited on the surface of the particles of $LiFePO_4$ has a greater conjugation.

The peaks at 796 $cm^{-1}$ assigned to N-H wag vibration and at 3229 $cm^{-1}$ assigned to N-H stretching vibration are not present. This is attributed to chemical interaction by the hydrogen bond N-H–O-P. The changes in the vibration values in the $LiFePO_4$–PANI compared with $LiFePO_4$ and PANI separately are attributed to chemical interaction by hydrogen bonds between the electronic clouds of PANI and $LiFePO_4$, which produce polarization and change in the dipolar moment.

The R value was calculated in order to find the ratio of the quinoid and benzenoid units (C = N/C-N). The ratio is directly related to the oxidation state of the PANI polymer. The value can be obtained by taking the ratio of the areas of the IR bands at ~1300 and ~1590 $cm^{-1}$ in Figure 11. The ratio R is calculated on the basis of Equation (1) [37]:

$$\frac{\nu\ \mathrm{N{=}C_6H_4{=}N}}{\nu\ \mathrm{N{-}C_6H_4{-}N}} = R = \frac{1-X}{X} \quad (1)$$

The values based on the areas determined from the FTIR data are R = 0.462 for PANI and R = 0.500 for $LiFePO_4$–PANI, respectively. When the ratio of the areas of the two bands is less than one, this indicates that there are more benzoid than quinoid units within the PANI polymer. Therefore, the structure of PANI and $LiFePO_4$–PANI doped with AcOLi can be represented as shown in Figure 12.

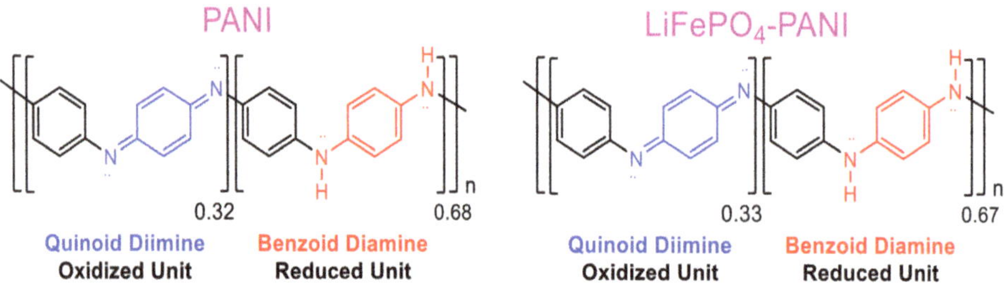


**Figure 12.** Oxidized and reduced units of PANI and $LiFePO_4$–PANI.

The ratio between quinoid and benzoid units in the PANI and $LiFePO_4$–PANI are similar. This indicates that the thermal treatment does not produce significant changes in the structure of the PANI. A possible structure and the interaction between the $LiFePO_4$ and PANI are shown in Figure 13.

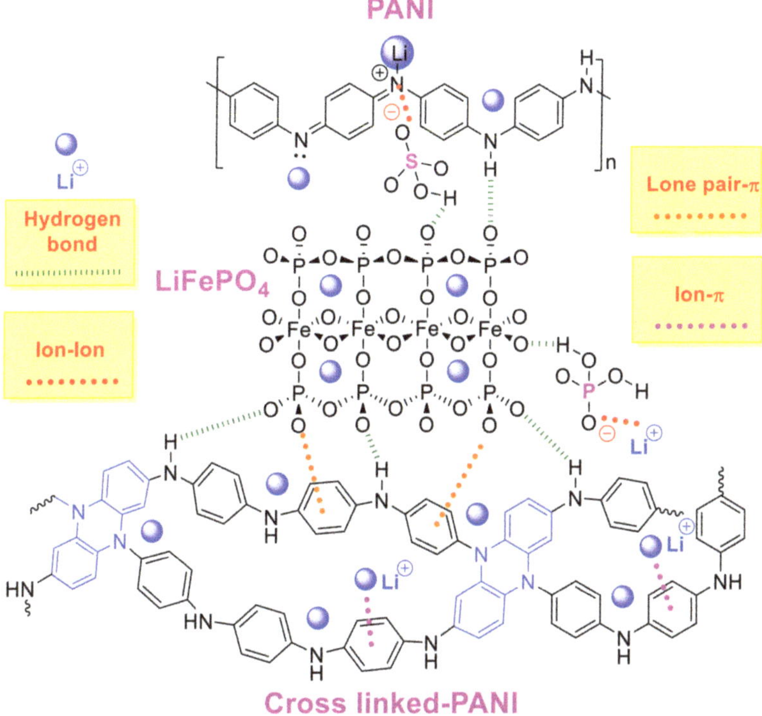


**Figure 13.** Possible structure of the $LiFePO_4$–PANI and its interactions.

Possible interactions present in the structure are hydrogen bonds, ion–ion, oxygen lone pair–π and ion–π. These interactions are weak and indicate the possibility of interaction and formation of a small amount of crosslinked PANI during the thermal treatment. A possible structure for PANI is shown in Supplementary Materials Figure S5.

### 3.5. Thermogravimetric Analysis

The TGA measurements of the PANI and $LiFePO_4$–PANI powders were carried out in air atmosphere, and the TGA curves are shown in Figure 14. The mass loss from room temperature to 150 °C is due to the elimination of small amounts of water absorbed in the $LiFePO_4$–PANI and PANI. The pristine PANI has a greater mass loss in this temperature interval and, thus, higher water content than the hybrid powder. This can be explained by the heat treatment at 300 °C in inert atmosphere during the synthesis, which dried the hybrid powder. The pristine PANI is completely decomposed between 150 and 600 °C. The PANI in $LiFePO_4$–PANI is decomposed at 500 °C. $LiFePO_4$ is still stable up to 850 °C, and the residual mass of carbon formed by decomposition of the polymer can be ignored. Thus, the mass content of the polymer can be calculated from the mass loss of the composite in the TGA curve. This calculated mass content of PANI in the $LiFePO_4$–PANI composites is 25%, which corresponds well to the precursor weights in the synthesis step. Possible decomposition reactions are shown in Supplementary Materials Figures S6 and S7.

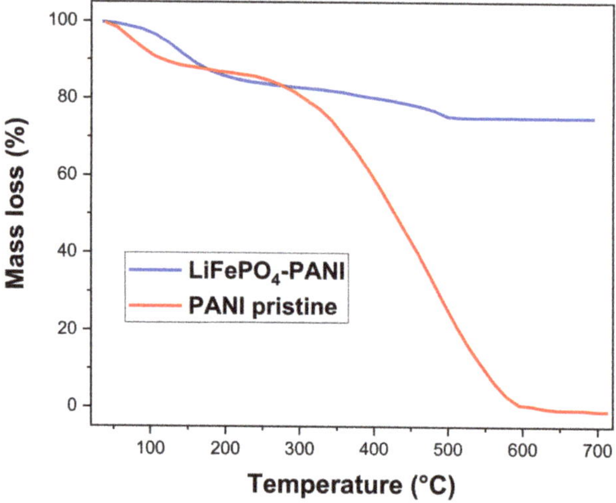


**Figure 14.** Thermogravimetric curves for pristine PANI and $LiFePO_4$–PANI.

### 3.6. Electrochemical Characterization

The $LiFePO_4$–PANI composites were electrochemically tested in a button-type test cell by charging and discharging at room temperature over 50 cycles, at a rate of 0.1 C. Figure 15 shows the charge–discharge capacity vs. cycle number for the $LiFePO_4$–PANI composite. The $LiFePO_4$ and $LiFePO_4$–PANI doped with $Li^+$ are electrochemically active in the range of 2.4–4.2 V. The $LiFePO_4$–PANI shows a small capacity fade of about 0.2% over 50 cycles.

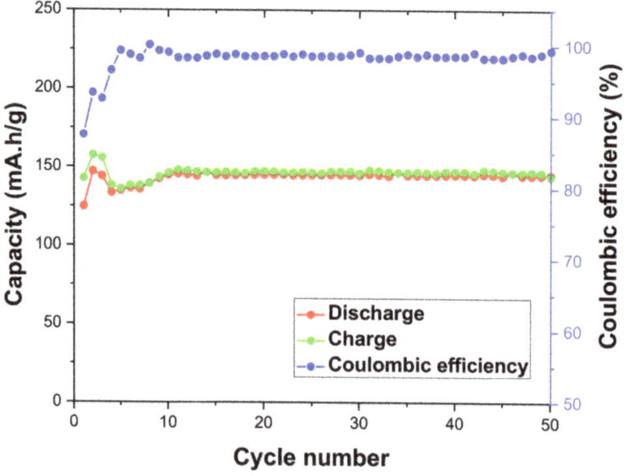


**Figure 15.** Charge/discharge cycling performances for the $LiFePO_4$–PANI composites. The data were obtained by charging and discharging at 0.1 C.

The cell voltage as a function of specific capacity for $LiFePO_4$–PANI composites tested at 2.4–4.2 V is shown in Figure 16. The inserted number of Li-ions determines the specific capacity of the materials, and the occupation sites of Li-ions control the electrochemical potential. $LiFePO_4$–PANI materials have one equivalent site in $LiFePO_4$, which suggests the same site energy for each Li-ion, resulting in a single continuous discharge plateau and a constant cell voltage. Supplementary Materials Figure S10 shows the charge and discharge profiles at 0.1 C as a function of time.

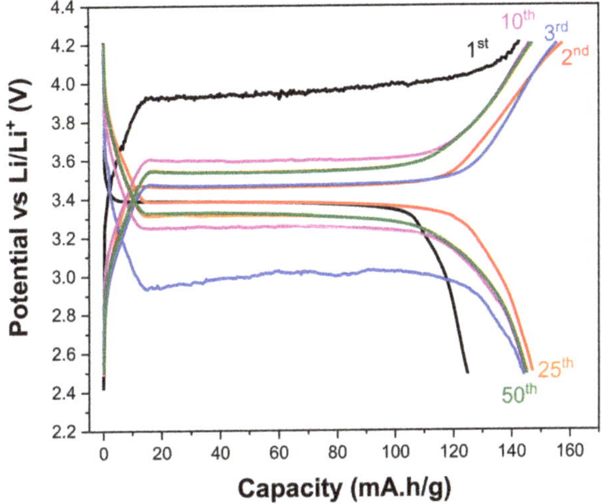


**Figure 16.** Cell voltage as a function of specific capacity obtained by charging and discharging at the same rate for the $LiFePO_4$–PANI. The data were obtained by charging and discharging at 0.1 C.

From Figure 17, the specific discharge capacity of $LiFePO_4$–PANI at 0.1 C is 145 mAh $g^{-1}$, and the corresponding result for pure $LiFePO_4$ is 120 mAh $g^{-1}$. The theoretical capacity of PANI is 142 mAh $g^1$, and that of $LiFePO_4$ is 170 mAh $g^{-1}$. The coulombic efficiency of the $LiFePO_4$–PANI reaches 98%. The $LiFePO_4$–PANI composite with 25 wt.% PANI exhibits a superior performance with a specific capacity of 145 mAh $g^{-1}$, compared to 120 mAh $g^{-1}$ for pure $LiFePO_4$ and 95 mAh $g^{-1}$ for pure PANI, as shown in Figure 17. This can be, in part, ascribed to the improved electrical conductivity of $LiFePO_4$ by its interaction with the PANI structure. The charge–discharge and cell voltage curves of the PANI are shown in Supplementary Materials Figures S8 and S9.

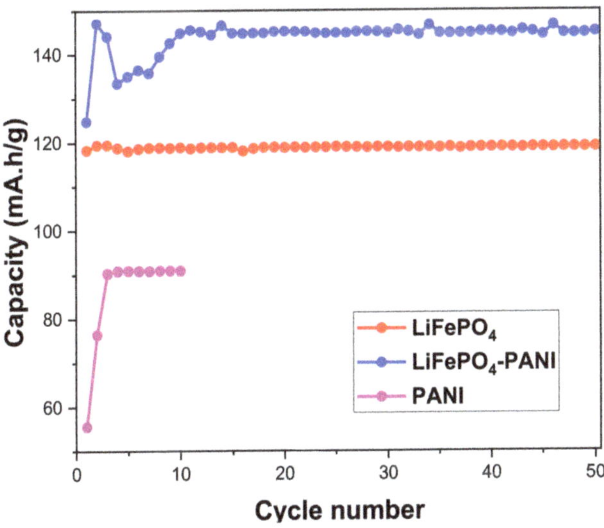


**Figure 17.** Comparison of discharge cycling performances for the $LiFePO_4$–PANI composite and $LiFePO_4$. The data were obtained by charging and discharging at 0.1 C.

PANI, $LiFePO_4$ and $LiFePO_4$–PANI cycled at different C-rates are compared in Figure 18. The $LiFePO_4$–PANI shows the best rate capability with discharge capacities of 145 mAh $g^{-1}$ at 0.1 C and 100 mAh $g^1$ at 2 C. This means that incorporation of PANI on the surface of the particles of $LiFePO_4$ results in 21% capacity enhancement at 0.1 C and 45% enhancement at 2 C. The enhanced rate capability is attributed to the improved electrical and ionic conductivity and $Li^+$ diffusion promoted by PANI at the surface of the $LiFePO_4$ particles.

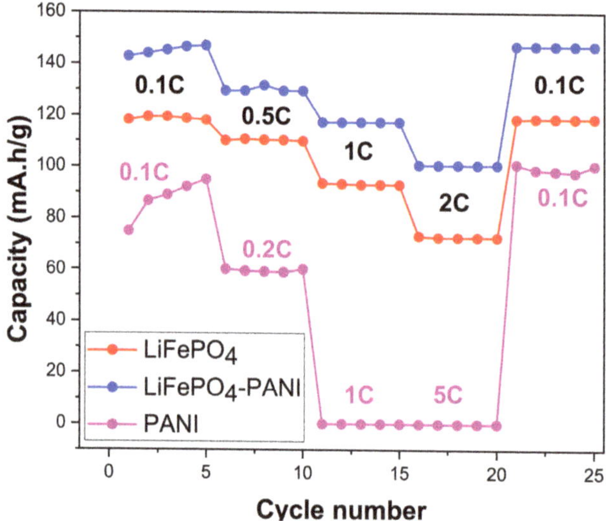


**Figure 18.** Specific discharge capacity at different C-rates for the $LiFePO_4$–PANI, PANI and pure $LiFePO_4$. The cells were charged and discharged at the same rates and conditions, from 0.1 C to 2 C.

The electrochemical impedance spectroscopy (EIS) curves for the PANI, $LiFePO_4$ and $LiFePO_4$–PANI composite cathodes at OCV (2.40 V for PANI, 2.17 V for $LiFePO_4$ and 3.42 V for $LiFePO_4$–PANI) are shown in Figure 19.

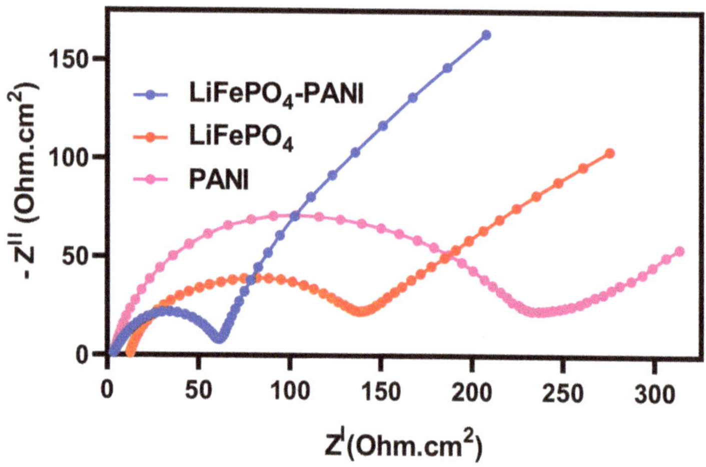


**Figure 19.** Experimental Nyquist curves for PANI, $LiFePO_4$ and $LiFePO_4$–PANI.

The high frequency real axis-intercept, $R_1$ in Table 1, represents the ionic bulk resistance and the electrical contact resistance, and the semicircle formed from the middle frequency range of the impedance spectra represents the charge transfer resistance, $R_{ct}$ [38,39]. The smaller diameter of the $LiFePO_4$–PANI semicircle shows a lower charge transfer resistance ($R_{ct}$) for the composite electrode, as shown in Table 1. This indicates that the $LiFePO_4$ coated or combined with conductive PANI can effectively improve the charge transfer and the electrochemical properties of $LiFePO_4$ particles.

**Table 1.** Data for impedance spectra at different cycling stages.

| Material | $R_1$ (Ohm) | $R_{ct}$ (Ohm) |
|---|---|---|
| $LiFePO_4$ as assembled (OCV) | 14.13 | 178 |
| $LiFePO_4$–PANI as assembled (OCV) | 3.58 | 61.1 |
| PANI as assembled (OCV) | 4.12 | 238 |

The results can be correlated with Figure 20, which shows the different processes in the proposed structure. The PANI also reduces the contact resistance by increasing the contact area between $LiFePO_4$ particles and the current collector. Figure 20 shows the electrochemical reaction of $LiFePO_4$ and the scheme of the $LiFePO_4$–PANI charging process. In the charge transfer process, the electron first moves through the molecular orbital network in the $LiFePO_4$. After that, the electron is transferred to the PANI chain and moves through the polymer chain to the current collector. The $Li^+$ moves through the network of the $LiFePO_4$ structure and the PANI structure to the electrolyte.

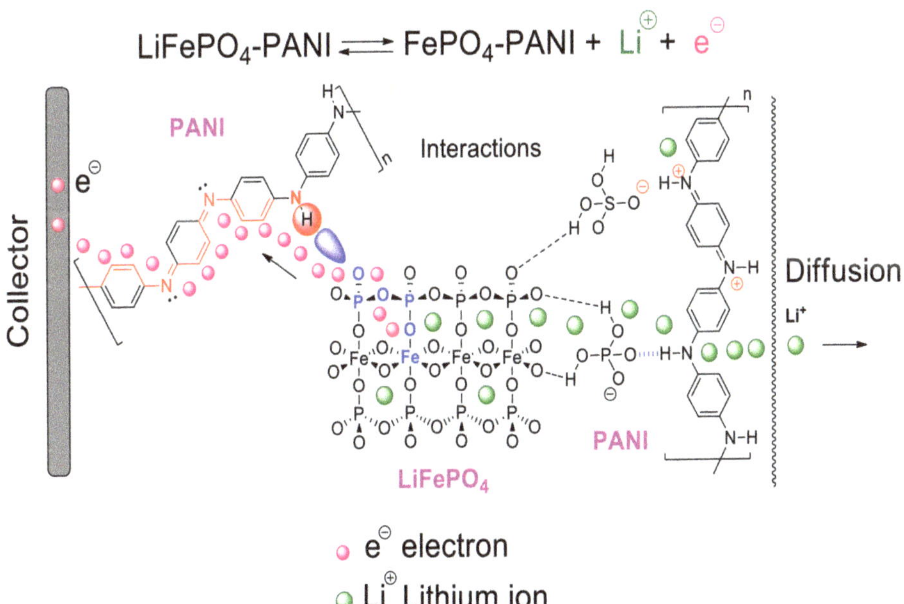


**Figure 20.** Scheme of the transfer and electron and diffusion of $Li^+$.

A qualitative structure of the $LiFePO_4$–PANI is shown in Figure 21. The surface $LiFePO_4$ particles are coated with PANI.

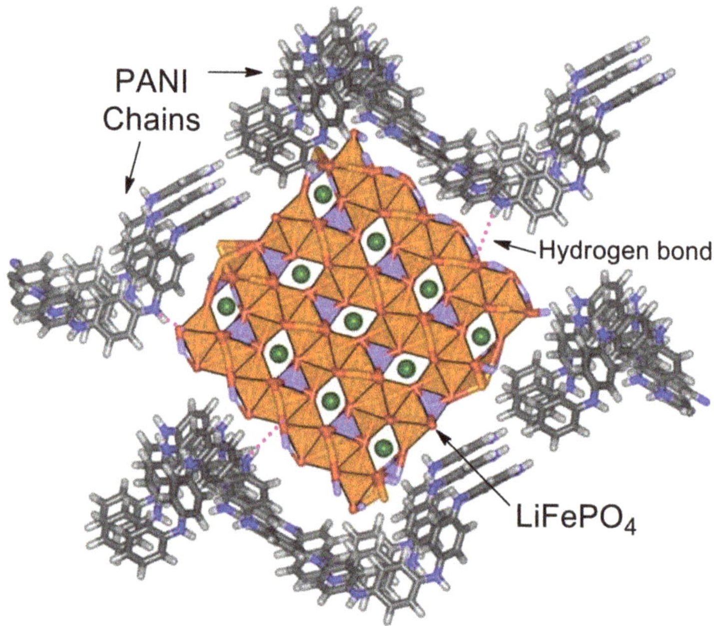


**Figure 21.** Qualitative structure for $LiFePO_4$–PANI.

The electrochemical results of the $LiFePO_4$–PANI reported in this study are similar to those of other works [12,40–44]. The specific capacity of $LiFePO_4$–PANI depends of the methods of the synthesis, dopants in PANI and the electrochemical properties of the $LiFePO_4$. All the works show an enhancement of the specific capacity at lower C-rates.

## 4. Conclusions

PANI was synthesized by chemical oxidation with good crystallinity, with primary particle size of 0.31 um and globular morphology forming agglomerates of 2.75 um. A $LiFePO_4$–PANI composite was prepared via a thermal process, using $LiFePO_4$, PANI and AcOLi. The morphologies of the $LiFePO_4$–PANI and $LiFePO_4$ particles are similar, but the PANI coating on the $LiFePO_4$ particles gives a different surface structure in the composite. This is supported by elemental mapping, which shows a homogeneous distribution of carbon and sulfur in PANI on the surface of $LiFePO_4$ particles.

The composite electrodes show good capacity, rate capability and cyclability. The composite containing 25 wt.% PANI shows better electrochemical performance compared with pure PANI and pure $LiFePO_4$. The discharge capacity of PANI is 95 mAh $g^{-1}$, that of $LiFePO_4$ is 120 mAh $g^{-1}$ and that of synthesized $LiFePO_4$–PANI is 145 mAh $g^{-1}$ at 0.1 C. At 2 C, the discharge capacity is 70 mAh $g^{-1}$ for $LiFePO_4$ and 100 mAh $g^{-1}$ for $LiFePO_4$–PANI. The PANI can be considered as a conductor for the electronic transfer between the $LiFePO_4$ particles, but it also improves the contact between the electrolyte and $LiFePO_4$ particles during the charge/discharge process. The PANI can also act as a host for $Li^+$ ion insertion–extraction and gives an additional contribution to the capacity of the composite. PANI can also serve as a binder network between $LiFePO_4$ particles and the current collector surface.

**Prime Novelty Statement:** In this work, we developed a new route to synthesize a $LiFePO_4$–PANI hybrid via thermal treatment, using lithium iron phosphate ($LiFePO_4$), polyaniline (PANI) and lithium acetate (AcOLi) as starting materials. The synthesis process has a high yield (99.99%). The structural, morphological and compositional properties of the $LiFePO_4$–PANI hybrid are characterized. Using the $LiFePO_4$–PANI hybrid as electrode materials for lithium-ion battery application, good electrochemical performance was obtained, showing a discharge capability of 145 mAh $g^{-1}$ at 0.1 C, 120 mAh $g^{-1}$ at 1C and 100 mAh $g^{-1}$ at 2 C, and cycling stability with 99.0% capacity retention at 0.1 C after 50 cycles. These initial research results are very interesting, and the technology developed in this work will provide a valuable approach to new potential cathode materials.

**Supplementary Materials:** The following are available online at http://www.mdpi.com/1996-1944/13/12/2834/s1. Figure S1: Mechanism of aniline polymerization in the synthesis PANI. Figure S2: Morphology of PANI, a, upper) illustration of primary particles and agglomerates, b, lower) SEM images of the synthesized PANI agglomerates and enlargement showing the globular morphology. Figure S3: EDS spectrum of the PANI. Figure S4. FTIR spectra for PANI. Figure S5. Possible structure of the PANI and its interactions. Figure S6. Possible qualitative reactions of the decomposition of the PANI during the TG analysis. Figure S7. Possible qualitative reactions of the decomposition of the $LiFePO_4$-PANI during the TG analysis. Figure S8. Electrochemical performance of PANI: cycling performance profiles. Figure S9. Electrochemical performance of PANI: galvanostatic discharge-charge profiles. Figure S10. Cell voltage as a function of time at different cycles obtained by charging and discharging at same rate for the $LiFePO_4$-PANI. The data were obtained by charging and discharging at 0.1 C.

**Author Contributions:** Conceptualization, methodology, software, validation and formal analysis, investigation, data curation, writing—Original draft preparation, and visualization, C.A.; methodology and investigation, N.L.; methodology, M.V.; writing—Review and supervision, G.L., A.L. and S.C.; project administration, G.L. and S.C.; funding acquisition, G.L. and S.C. All authors have read and agreed to the published version of the manuscript.

**Funding:** This research was funded by SIDA (Swedish International Development Agency); KTH Royal Institute of Technology, Department of Chemical Engineering, Applied Electrochemistry; UMSA (Universidad Mayor de San Andres), Department of Inorganic Chemistry and Materials Science/Advanced Materials, and IIQ Chemical Research Institute.

**Acknowledgments:** The authors greatly acknowledge the financial support by: SIDA (Swedish International Development Agency); KTH Royal Institute of Technology, Department of Chemical Engineering, Applied Electrochemistry; UMSA (Universidad Mayor de San Andres), Department of Inorganic Chemistry and Materials Science/Advanced Materials, IIQ Chemical Research Institute.

**Conflicts of Interest:** The authors declare no conflict of interest.

## References

1. Padhi, A.K.; Nanjundaswamy, K.S.; Goodenough, J.B. Phospho-olivines as Positive-Electrode Materials for Rechargeable Lithium Batteries. *J. Electrochem. Soc.* **1997**, *144*, 1188–1194. [CrossRef]
2. Ajpi, C.; Diaz, G.; Visbal, H.; Hirao, K. Synthesis and characterization of Cu-doped $LiFePO_4$ with/without carbon coating for cathode of lithium-ion batteries. *J. Ceram. Soc. Jpn.* **2013**, *121*, 441–443. [CrossRef]
3. Lin, H.B.; Zhang, Y.M.; Hua, J.N.; Wang, Y.T.; Xing, L.; Xu, M.Q.; Li, X.P.; Li, W.S. $LiNi_{0.5}Mn_{1.5}O_4$ nanoparticles: Synthesis with synergistic effect of polyvinylpyrrolidone and ethylene glycol and performance as cathode of lithium ion battery. *J. Power Sources* **2014**, *257*, 37–44. [CrossRef]
4. Villca, J.; Vargas, M.; Yapu, W.; Blanco, M.; Benavente, F.; Cabrera, S. Nueva ruta pirolitico/atrano para la síntesis de electrodos catódicos $LiCoO_2$ y $LiMn_2O_4$. *Revista Boliviana de Química* **2014**, *31*, 82–85.
5. Lee, M.; Hong, J.; Kim, H.; Lim, H.-D.; Cho, S.B.; Kang, K.; Park, C.B. Organic Nanohybrids for Fast and Sustainable Energy Storage. *Adv. Mater.* **2014**, *26*, 2558–2565. [CrossRef] [PubMed]
6. Morales, J.; Trocoli, R.; Rodriguez-Castellon, E.; Franger, S.; Santos-Pena, J. Effect of C and Au additives produced by simple coaters on the surface and the electrochemical properties of nanosized $LiFePO_4$. *J. Electroanal. Chem.* **2009**, *631*, 29–35. [CrossRef]
7. Chung, S.Y.; Bloking, J.T.; Chiang, Y.M. Electronically conductive phospho-olivines as lithium storage electrodes. *Nat. Mater.* **2002**, *1*, 123–128. [CrossRef]
8. Hiu, H.; Cao, Q.; Fu, L.J.; Li, C.; Wu, Y.P.; Wu, H.Q. Doping effects of zinc on $LiFePO_4$ cathode material for lithium ion batteries. *Electrochem. Commun.* **2006**, *8*, 1553–1557.
9. Prosini, P.P.; Carewska, M.; Scaccia, S.; Wisniewski, P.; Passerini, S. A new synthetic route for preparing $LiFePO_4$ with enhanced electrochemical performance. *J. Electrochem. Soc.* **2002**, *149*, A886–A890. [CrossRef]
10. Bruce, P.G.; Scrosati, B.; Tarascon, J.-M. Nanomaterials for rechargeable lithium batteries. *Angew. Chem. Int.* **2008**, *47*, 2930–2946. [CrossRef]
11. Park, K.S.; Schougaard, S.B.; Goodenough, J.B. Conducting-Polymer/Iron-Redox-Couple Composite Cathodes for Lithium Secondary Batteries. *Adv. Mater.* **2007**, *19*, 848–851. [CrossRef]
12. Chen, W.-M.; Huang, Y.-H.; Yuan, L.-X. Self-assembly $LiFePO_4$/polyaniline composite cathode materials with inorganic acids as dopants for lithium-ion batteries. *J. Electroanal. Chem.* **2011**, *660*, 108–113. [CrossRef]
13. Abdiryim, T.; Zhang, X.-G.; Jamal, R. Comparative studies of solid-state synthesized polyaniline doped with inorganic acids. *Mater. Chem. Phys.* **2005**, *90*, 367–372. [CrossRef]

14. Sisbandini, C.; Brandell, D.; Gustafsson, T.; Nyholm, L. The Mechanism of Capacity Enhancement in $LiFePO_4$ Cathodes through Polyetheramine Coating. *J. Electrochem. Soc.* **2009**, *156*, A720–A725. [CrossRef]
15. Ayad, M.; Zaghlol, S. Nanostructured crosslinked polyaniline with high surface area: Synthesis, characterization and adsorption for organic dye. *Chem. Eng. J.* **2012**, *204–206*, 79–86. [CrossRef]
16. Chatterjee, M.J.; Ghosh, A.; Mondal, A.; Banerjee, D. Polyaniline–single walled carbon nanotube composite—A photocatalyst to degrade rose bengal and methyl orange dyes under visible-light illumination. *RSC Adv.* **2017**, *7*, 36403–36415. [CrossRef]
17. Huang, J.X.; Virji, S.; Weiller, B.H.; Kaner, R.B. Polyaniline nanofibers: Facile synthesis and chemical sensors. *J. Am. Chem. Soc.* **2003**, *125*, 314. [CrossRef]
18. Zhang, Z.; Wan, M.; Wei, Y. Highly Crystalline Polyaniline Nanostructures Doped with Dicarboxylic Acids. *Adv. Funct. Mater.* **2006**, *16*, 1100–1104. [CrossRef]
19. Fonseca, C.P.; Neves, S. The usefulness of a $LiMn_2O_4$ composite as an active cathode material in lithium batteries. *J. Power Sources* **2004**, *135*, 249. [CrossRef]
20. Wang, Y.G.; Wu, W.; Cheng, L.; He, P.; Wang, C.X.Y.; Xia, Y. A polyaniline-intercalated layered manganese oxide nanocomposite prepared by an inorganic/organic interface reaction and its high electrochemical performance for Li storage. *Adv. Mater.* **2008**, *20*, 2166. [CrossRef]
21. Kang, S.-G.; Man Kim, K.; Park, N.-G.; Sun Ryu, K.; Chang, S.-H. Factors affecting the electrochemical performance of organic/$V_2O_5$ hybrid cathode materials. *J. Power Sources* **2004**, *133*, 263–267. [CrossRef]
22. Karthikeyan, K.; Amaresh, S.; Aravindan, V.; Kim, W.S.; Namd, K.W.; Yang, X.Q.; Lee, Y.S. $Li(Mn_{1/3}Ni_{1/3}Fe_{1/3})O_2$ Polyaniline hybrids as cathode active material with ultra-fast chargeedischarge capability for lithium batteries. *J. Power Sources* **2013**, *232*, 240–245. [CrossRef]
23. Novák, P.; Müller, K.; Santhanam, K.S.V.; Haas, O. Electrochemically active polymers for rechargeable batteries. *Chem. Rev.* **1997**, *97*, 207.
24. Chen, S.A.; Lin, L.C. Polyaniline Doped by the New Class of Dopant, Ionic Salt: Structure and Properties. *Macromolecules* **1995**, *28*, 1239. [CrossRef]
25. Zhang, Z.; Wan, M. Nanostructures of polyaniline composites containing nano-magnet. *Synth. Mater.* **2003**, *132*, 205–212. [CrossRef]
26. Kim, B.K.; Kim, Y.H.; Won, K.; Chang, H.; Choi, Y.; Kong, K.; Rhyu, B.W.; Kim, J.J.; Lee, J.O. Electrical properties of polyaniline nanofibre synthesized with biocatalyst. *Nanotechnology* **2005**, *16*, 1177–1181. [CrossRef]
27. Ding, H.; Long, Y.; Shen, J.; Wan, M.; Wei, Y. $Fe_2(SO_4)_3$ as a binary oxidant and dopant to thin polyaniline nanowires with high conductivity. *J. Phys. Chem. B* **2010**, *114*, 115–119. [CrossRef]
28. Zhang, Z.; Wei, Z.; Wan, M. Nanostructures of polyaniline doped with inorganic acids. *Macromolecules* **2002**, *35*, 5937–5942. [CrossRef]
29. Sanchez, C.; Rozes, L.; Ribot, F.; Laberty-Robert, C.; Grosso, D.; Sassoye, C.; Boissiere, C.; Nicole, L. Chimie douce: A land of opportunities for the designed construction of functional inorganic and hybrid organic-inorganic nanomaterials. *Comptes Rendus Chim.* **2010**, *13*, 3–39. [CrossRef]
30. Mostafaei, A.; Zolriasatein, A. Synthesis and characterization of conducting polyaniline nanocomposites containing ZnO nanorods. *Prog. Nat. Sci. Mater. Int.* **2012**, *22*, 273–280. [CrossRef]
31. Pouget, J.P.; Jozefowicz, M.E.; Epstein, A.J.; Tang, A.G. X-ray structure of Polyaniline. *Macromolecules* **1991**, *24*, 779–789. [CrossRef]
32. Paques-Ledent, M.T.; Tarte, P. Vibrational studies of olivine-typecompounds II orthophosphates, arsenates and vanadates $A^IB^{II}{}_XVO_4$. *Spectrochim. Acta Part A Mol. Spectrosc.* **1974**, *30*, 673–689. [CrossRef]
33. Li, W.; Hwanga, J.; Chang, W.; Setiadi, H.; Chung, K.Y.; Kim, J. Ultrathin and uniform carbon-layer-coated hierarchically porous $LiFePO_4$ microspheres and their electrochemical performance. *J. Supercrit. Fluids* **2016**, *116*, 164–171. [CrossRef]
34. Salah, A.A.; Jozwiak, P.; Zaghib, K.; Garbarczyk, J.; Gendron, F.; Mauger, A.; Julien, C. FTIR features of lithium-iron phosphates as electrode materials for rechargeable lithium batteries. *Spectrochim. Acta Part. A* **2006**, *65*, 1007–1013. [CrossRef]
35. Neoh, K.G.; Kang, E.T.; Tan, K.L. Evolution of polyaniline structure during synthesis. *Polymer* **1993**, *34*, 3921–3928. [CrossRef]

36. Nagaraja, M.; Mahesh, H.M.; Manjanna, J.; Rajanna, K.; Kurian, M.Z.; Lokesh, S.V. Effect of Multiwall Carbon Nanotubes on Electrical and Structural Properties of Polyaniline. *J. Electron. Mater.* **2012**, *41*, 1882–1885. [CrossRef]
37. Hatchett, D.W.; Josowicz, M.; Janata, J. Acid Doping of Polyaniline: Spectroscopic and Electrochemical Studies. *J. Phys. Chem. B* **1999**, *103*, 10992–10998. [CrossRef]
38. Tian, B.B.; Ning, G.-H.; Gao, Q.; Tan, L.M.; Tang, W.; Chen, Z.; Su, C.; Loh, K.P. Crystal Engineering of Naphthalenediimide-Based Metal-Organic Frameworks: Structure-Dependent Lithium Storage. *ACS Appl. Mater. Interfaces* **2016**, *8*, 31067–31075. [CrossRef]
39. Shi, Y.; Shen, M.; Xu, S.; Qiu, X.; Jiang, L.; Qiang, Y.-H.; Zhuang, Q.C.; Sun, S. Electrochemical Impedance Spectroscopic Study of the Electronic and Ionic Transport Properties of $NiF_2$/C Composites. *Int. J. Electrochem.* **2011**, *6*, 3399–3415.
40. Fagundesa, W.S.; Xaviera, F.F.S.; Santanaa, L.K.; Azevedoa, M.E.; Canobrea, S.C.; Amarala, F.A. PAni-coated $LiFePO_4$ Synthesized by a Low Temperature Solvothermal Method. *Mater. Res.* **2019**, *22*, e20180566. [CrossRef]
41. Gong, W.; Chen, Y.; Sun, B.; Chen, H. Electrochemical performance of $LiFePO_4$ coated by nano-polyaniline coater for lithium-ion batteries. *Adv. Mat. Res.* **2010**, *146*, 551–555.
42. Bai, Y.-M.; Qiu, P.; Wen, Z.-L.; Han, S.-C. Synthesis and electrochemical properties of $LiFePO_4$/PANI composites. *Chin. J. Nonferr. Met.* **2010**, *28*, 8.
43. Gong, C.; Deng, F.; Tsui, C.-P.; Xue, Z.; Ye, Y.S.; Tang, C.-Y.; Zhou, X.; Xie, X. PANI–PEG copolymer modified $LiFePO_4$ as a cathode material for high-performance lithium ion batteries. *J. Mater. Chem. A* **2014**, *2*, 19315. [CrossRef]
44. Gong, Q.; He, Y.-S.; Yang, Y.; Liao, X.-Z.; Ma, Z.-F. Synthesis and electrochemical characterization of $LiFePO_4$/C-polypyrrole composite prepared by simple chemical vapor deposition method. *J. Solid State Electrochem.* **2012**, *16*, 1383–1388. [CrossRef]

 *materials*

*Article*

# Optical and Electrochemical Applications of Li-Doped NiO Nanostructures Synthesized via Facile Microwave Technique

**Aarti S. Bhatt [1], R. Ranjitha [2], M. S. Santosh [3,*], C. R. Ravikumar [4,*], S. C. Prashantha [4], Rapela R. Maphanga [5,6] and Guilherme F. B. Lenz e Silva [7]**

1 Department of Chemistry, N.M.A.M. Institute of Technology (Visvesvaraya Technological University, Belagavi), Nitte 574110, India; aartis@nitte.edu.in
2 Department of Chemistry, St. Aloysius College (Autonomous), Mangaluru 575003, India; ranjut85@gmail.com
3 Centre for Incubation, Innovation, Research and Consultancy (CIIRC), Jyothy Institute of Technology, Thataguni, Off Kanakpura Road, Bangalore 560082, Karnataka, India
4 Research Centre, Department of Chemistry, East West Institute of Technology, Bengaluru 560091, India; scphysics@gmail.com
5 Next Generation Enterprises and Institutions, Council for Scientific and Industrial Research, Pretoria 0001, South Africa; RMaphanga@csir.co.za
6 Department of Physics, University of Limpopo, Private bag x 1106, Sovenga 0727, South Africa
7 Polytechnic School of Engineering, University of São Paulo, São Paulo 05508-030, Brazil; guilhermelenz@usp.br
* Correspondence: santoshgulwadi@gmail.com (M.S.S.); ravikumarcr@ewit.edu (C.R.R.)

Received: 21 January 2020; Accepted: 3 April 2020; Published: 2 July 2020

**Abstract:** Nanostructured NiO and Li-ion doped NiO have been synthesized via a facile microwave technique and simulated using the first principle method. The effects of microwaves on the morphology of the nanostructures have been studied by Field Emission Spectroscopy. X-ray diffraction studies confirm the nanosize of the particles and favoured orientations along the (111), (200) and (220) planes revealing the cubic structure. The optical band gap decreases from 3.3 eV (pure NiO) to 3.17 eV (NiO doped with 1% Li). Further, computational simulations have been performed to understand the optical behaviour of the synthesized nanoparticles. The optical properties of the doped materials exhibit violet, blue and green emissions, as evaluated using photoluminescence (PL) spectroscopy. In the presence of Li-ions, NiO nanoparticles exhibit enhanced electrical capacities and better cyclability. Cyclic voltammetry (CV) and electrochemical impedance spectroscopy (EIS) results show that with 1% Li as dopant, there is a marked improvement in the reversibility and the conductance value of NiO. The results are encouraging as the synthesized nanoparticles stand a better chance of being used as an active material for electrochromic, electro-optic and supercapacitor applications.

**Keywords:** Li-doped NiO; microwave synthesis; computational simulation; electrochemical measurements; photoluminescence

## 1. Introduction

Nano metal oxides are attracting several researchers due to their varied applications. The reduction in their size contributes tremendously to expanding the surface area, thereby making their optical and electrical properties highly sensitive to surface morphology. Among the various nanometal oxides, nickel oxide (NiO) nanoparticles have been extensively investigated because of their excellent chemical stability and favourable opto-electrical properties. NiO is an antiferromagnetic material and possesses a cubic structure. It has a band gap in the range of 3.6–3.8 eV [1] which can be tailored by reducing the size and/or by doping. It is therefore not surprising that NiO finds applications as antiferromagnetic materials [2], p-type transparent conducting films [3], electrochromic display materials [4] and chemical sensors [5].

For electro-optic applications, the stoichiometry of the crystal and optimization of the band structure are crucial. The chemical composition and crystal structure of the nanoparticles can be manipulated by introducing impurity atoms. The presence of dopants alters the energy configuration of the crystal lattice, thereby influencing the optical and transport properties. Generally, dopants from I to V group elements are introduced to obtain a stable p-type NiO semiconductor [6]. It has been observed that the presence of Li in the metal oxide lattice increases the electrical resistivity, thereby making them suitable for manufacturing transparent conducting oxides, piezoelectric devices and memory devices [7,8]. However, it is also claimed [9–11] that the presence of Li as a dopant results in a decrease in the resistivity of NiO leading to an improvement in its electrochromic properties. Hence, it becomes necessary to understand the effect of dopant concentration on the electrochemical properties.

The optical properties of the materials also play a crucial role in several applications, like optoelectronics, integrated optics, solar power engineering and optical sensor technology. Generally, compounds like $C_2S$, $Ca_2SiO_4$ and $CaAl_2O_4$ are considered as standard phosphor materials [12–15]. It is reported that doping of P ions in the green emitting $C_2S{:}Eu^{2+}$ effectively enhances the PL intensities [13]. Doping these with rare earth metals like Eu and Nd is known to enhance their emission characteristics. Nakano et al. have reported the effect of annealing temperature on the photoluminescence of Eu doped $Ca_2SiO_4$. They observed a red emission at 1773 K and green emission at 1473 K [14]. Recently, a significant work on Eu/Nd-doped $Ca_{12}Al_{14}O_{33}$ and $CaAl_2O_4$ phosphors as long-lasting blue emitters has also been carried out [15]. Though binary metal oxides are well known for their superior luminous efficiency, related works on simple oxides are rare. Wide gap semiconductors like ZnO have been used extensively as nanophosphors [16,17]. It was found that the doping of these nanophosphors with dopants like Li [18], Al [19], Mg [20], V [21], Eu [22] and Er [23] significantly alters their visible emission spectra. It is interesting to note that the number of research papers reporting on the electrochemical properties of NiO materials are much greater in comparison to the reports on their optical properties. NiO, being a p-type semi-conductor, stands a fair chance of being employed as a light-emitting diode or a photodetector in photoelectronic devices. To the best of our knowledge, there are very few studies [18] exploring the effect of Li ion on the optical properties of NiO. The present work is an attempt to underline the effect of Li in tuning the optical behavior of NiO along with its electrochemical characteristics.

In this background, the present work is focused on understanding the influence of Li doping on the structural, optical and electrochemical properties of NiO. The nanoparticles were synthesized by microwave-assisted synthesis. Since here, high-frequency microwaves are used for heating, this method has an advantage over conventional heating of being faster and consuming low energy. Although there is literature available on the microwave synthesis of NiO [24–26] and doped NiO [27], to the best of our knowledge, the present work is the first attempt to synthesize Li-doped NiO by microwave radiations. A detailed mechanism has also been proposed to understand the decrease in the band gap and resistivity of the as-synthesized nanomaterials on inclusion of the dopant.

## 2. Materials and Methods

### 2.1. Synthesis of NiO Nanoparticles

Nickel acetate tetrahydrate, lithium carbonate, polyethylene glycol 200 (PEG 200) and sodium hydroxide pellets were obtained from Sigma Aldrich, Bangalore, India. All chemicals used in this work are of analytical grade and were used without any further purification. An aqueous solution of 0.2 M nickel acetate was prepared. To this, 0.3 M NaOH solution was added dropwise with continuous stirring. The mixture was stirred for an hour to achieve concentration homogeneity. The resulting green-colored solution was heated in a microwave oven for 5 min at 210 W and 95 °C. The green-colored product thus obtained was alternately washed with water and ethanol till a pH of 7.0 was attained. The precipitate was separated by centrifugation and the solid precursor obtained was annealed in a muffle furnace at 300 °C for 2 h.

In an alternate procedure, PEG 200 which acts as a surfactant, was dissolved in NaOH solution before adding it dropwise to nickel solution in a similar manner described earlier. The surfactant to $Ni(CH_3COO)_2$ ratio was fixed to 2:1. It can be inferred that the surfactant acts on the surface of NiO nuclei, thereby inhibiting its growth (Figure 1). In a similar manner, doped samples were prepared by adding lithium carbonate solutions of different concentrations (1, 2 and 5%) to nickel acetate solutions, followed by alkali addition. The overall idea of the synthetic method adapted is depicted in the flow chart represented in Figure 2.

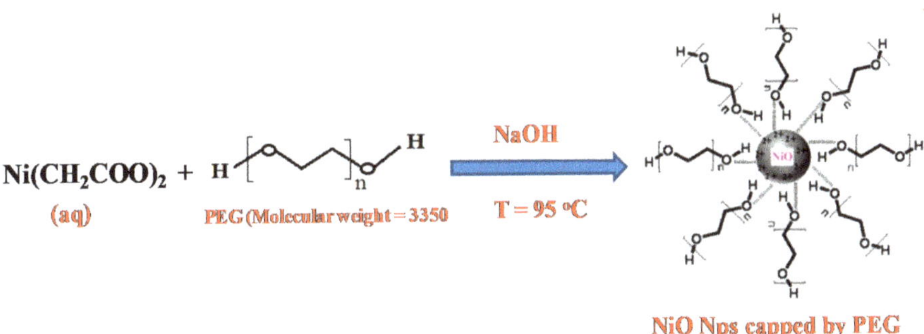


**Figure 1.** Structural impact of nickel acetate and polyethylene glycol(PEG) reaction.

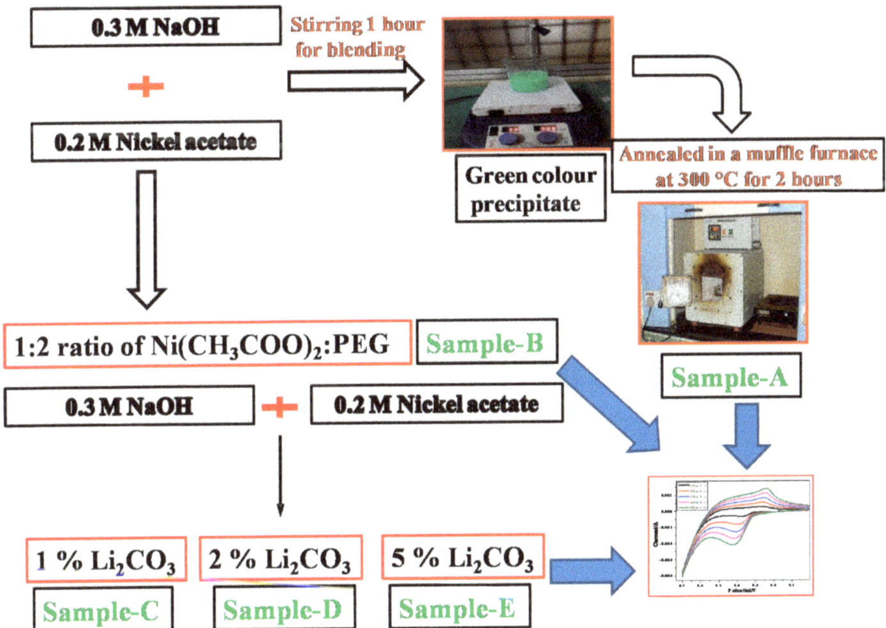


**Figure 2.** Flow chart for the synthesis of different NiO nanoparticles.

### 2.2. Characterization

X-ray diffraction studies were carried out on Philips X'pert PRO X-ray diffractometer, Malvern, UK, with graphite monochromatized $CuK_{\alpha}$ ($\lambda$ = 1.5418 Å) radiation, at a scan rate of 10° $min^{-1}$. Surface morphology was studied using a Field Emission Scanning Electron Microscopy (FESEM; Neon 40 Crossbeam, Carl Zeiss, Jena, Germany) with a resolution of 1.1 nm. Diffuse reflectance spectra were examined using the Shimadzu UV-VIS spectrophotometer (UV-2600, Kyoto City, Japan), in the range of 200–800 nm. Photoluminescence studies were carried out on a Horiba (model Fluorolog-3) spectrofluorimeter (Kyoto City, Japan) with 450 W Xenon excitation source. The spectral analysis was carried out using Fluor Essence™, Version 3.9.

### 2.3. Preparation of Carbon Paste Electrode for Electrochemical Studies

Cyclic voltammetry (CV) and electrochemical impedance spectroscopy (EIS) were performed on Model CHI604E potentiostat (CH Instruments, TX, USA). The experiment was carried out in acell comprised of a three-electrode system—a carbon paste working electrode (3.0 mm in diameter), a platinum wire counter electrode and a saturated Ag/AgCl reference electrode. The carbon paste electrode was prepared by grinding 70% graphite powder (particle size 50 µm and density 20 mg/100 mL), 15% of the prepared sample (A to E in Figure 2) and 15% silicone oil. The resulting homogeneous product was packed into the cavity of a customized polymer tube and smoothened at the ends.

### 2.4. Computational Details

The total energies were calculated by the projected augmented planewave (PAW) of density function-al theory, as implemented in the CASTEP code embedded in the Materials Studio software, Version 5.5.2 [28]. The core electrons were described using the projector augmented method. The exchange–correlation energy of the electrons was treated using the generalized gradient approximation within the Perdew–Burke–Ernzerhof functional [29]. For geometry optimization, the Monkhorst–Pack scheme was used. Ground state geometries were calculated by minimizing stresses and Hellman–Feynman forces using the conjugate gradient algorithm with the convergence parameters set as follows: total energy tolerance $2 \times 10^{-6}$ eV/atom, maximum force tolerance 0.05 eV/nm and maximum stress component 0.1 GPa. From the convergence test calculations, the basis set kinetic energy cutoff of 600 eV was sufficient to ensure optimum accuracy in the computed results. The dimensions for nanoclusters were set 20.00 × 20.00 × 20.00 Å, large enough to ensure that there were no interactions between the system and its self-image along all the axes within the periodic boundary conditions. Different sizes of the nanoparticles were constructed from the optimized bulk NiO system using supercells of varying sizes. The NiO nanoparticle was constructed by cleaving the bulk system along the (110) plane. Afterward, the Li ions were inserted into the nanoparticles using two different doping mechanisms, i.e., interstitial and substitutional doping. The k-points were generated using the Monkhorst-Pack method with a grid size of 1 × 1 × 1 for structural optimization for nanoparticle. The stoichiometry was maintained throughout nanoparticle construction and the vacuum was included in the x, y and z directions. The vacuum thickness was considered wide enough to prevent nanoparticle-to-nanoparticle interactions and 20 Å was sufficient to ensure that the energy converged to less than 1 meV/atom.

## 3. Results and Discussion

### 3.1. XRD Analysis

X-ray diffraction analysis was carried out to investigate the structures of pure NiO and Li doped NiO samples. Figure 3 shows the XRD peaks of pure NiO and samples prepared with (Sample B) and without (Sample A) surfactant, along with Li doped NiO (Samples C, D and E). The XRD patterns reveal the polycrystalline nature of the samples. The observed $2\theta$ values are in good agreement with the standard JCPDS data (Card ID 75-0197). All the peaks are well indexed, with the favoured orientations being (111), (200) and (220) planes [23]. From the analysis of the peak positions and comparative intensities of the diffracted peaks, it was confirmed that the samples were monophasic, Fm-3m cubic NiO. Since the sample is in powder form, Scherrer equation (Equation (1)) was used to relate the size of the crystallites to the broadening of the peak. The crystallite sizes, evaluated using the (200) peak as reference, are tabulated in Table 1.

$$\tau = {}^{K\lambda}/_{\beta cos\theta} \tag{1}$$

where $\tau$ is the mean size, $K$ is a dimensionless shape factor with a value of 0.94, $\lambda$ is X-ray wavelength, $\beta$ is the FWHM value and $\theta$ is the Bragg angle. From Table 1, it can be inferred that the presence of PEG during the synthesis of nanoparticles inhibits grain growth, as manifested by the peak broadening. PEG plays a crucial role in modifying the surface properties, thereby arresting grain growth. It is also observed that the Li doping of NiO also modified its crystallite size. This may be mainly due to the slightly larger size and lower charge of $Li^+$ compared to that of the $Ni^{2+}$; ionic radii of $Li^+$ and $Ni^{2+}$, which are 0.74 Å and 0.69 Å, respectively [30]. This suggests that, upon doping with Li, the dopant ions enter the rock salt crystal lattice of NiO substitutionally, and the surface stress may retard the growth of NiO nanomaterials. However, what is curious is that, at larger dopant concentrations (at 5%), this behaviour is not so pronounced. As the concentration of dopant increases, the peak intensity increases, revealing the higher crystallinity of the samples. This is reflected in the gradual increase in the crystallite size of the doped samples. Because of the low activation energy and high ionic mobility, the Li ions enter the nucleation sites, leading to increased strain and grain size [31]. A similar trend was observed by Matsubara et al. while doping NiO with Li up to 15 wt.%, who attributed this to the small amount of lattice contraction of NiO matrix on efficient substitution of $Li^+$ for $Ni^{2+}$ [32].

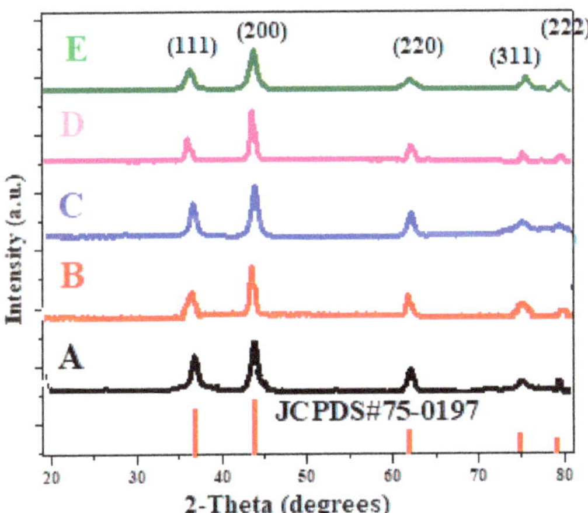


**Figure 3.** XRD pattern of A–E samples.

**Table 1.** Crystallite size calculated using X-ray diffractometer.

| Sample | Size (nm) |
|---|---|
| A | 3.08 |
| B | 2.47 |
| C | 2.36 |
| D | 2.43 |
| E | 2.80 |

### 3.2. Surface Morphology

Figure 4 shows the FESEM images of pure NiO and Li-doped samples with different concentrations of Li. It is clear that the NiO particles are of spherical shape with a size distribution in the 50–100 nm range. It is evident that the obtained NiO particles are porous in nature. This suggests that since, during the synthesis, a large amount of gases evolve, this results in the disintegration of agglomerates and non-uniform dissipation of heat within the system. This can hinder the grain growth and eventually lead to an increase in the surface area and porosity of the material. When compared to the undoped samples, the doped samples exhibit a more uniform micrograph. For samples C, D and E, the small crystallite size leads to agglomeration. However, as the dopant concentration increases, the particles display a more prominent spherical shape.

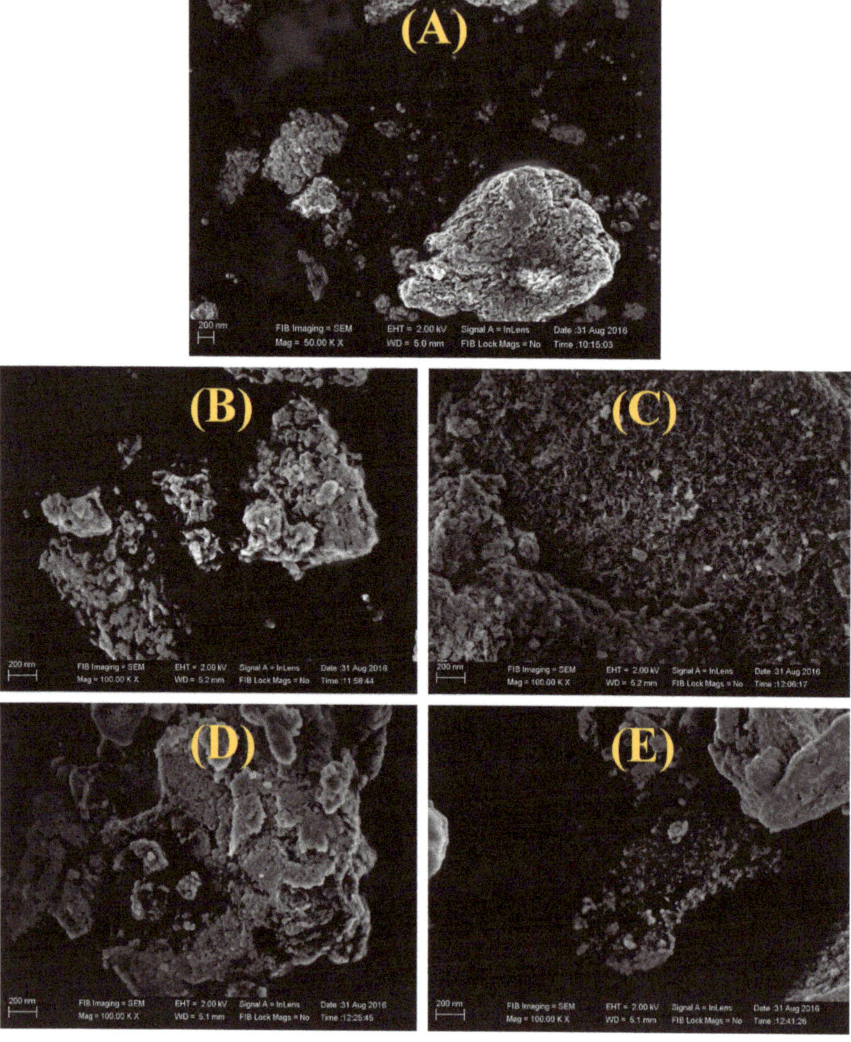


**Figure 4.** SEM images of (**A**) pristine NiO, (**B**) pristine NiO with PEG, (**C**) NiO with PEG and 1 wt% Li, (**D**) NiO with PEG and 2 wt% Li, (**E**) NiO with PEG and 5 wt% Li.

### 3.3. Computational Studies

Geometrically, NiO is a stacked rocksalt structure with a space group Fm-3m and lattice parameter 4.17 Å [33]. Three possible surface orientations for NiO can be obtained, which are (100), (110) and (111). The (100) is the most stable surface, with the surface energy of 1.15 J/$m^2$ [34]. The constructed nanoparticles for NiO and Li-inserted NiO are depicted in Figure 5 and have an octahedron geometry with a diameter of 3 nm. All nanoparticles possess a high symmetry, and their surfaces are characterized by (100) Miller index. A recent study investigating the effects of NiO nanoparticle surface energies on catalytic efficiency showed that (110) and (111) dominate the nanoparticle surface [35]. The Li doping of NiO has an impact on its crystal structure and also crystal size.

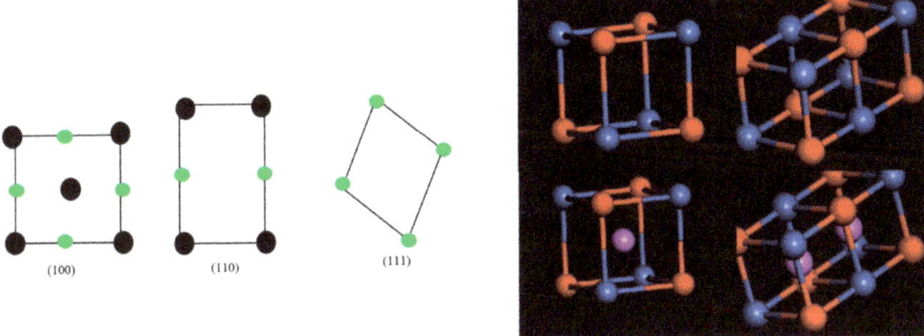


**Figure 5.** Constructed nanoparticles for NiO and Li-inserted NiO.

Computational modelling of optical functions for solids gives insights into understanding the optical properties of different materials. Due to the limitations and failure of GGA to capture the properties of strongly correlated electronic systems such as antiferromagnetic NiO, the simulated energy band gaps for the nanoparticles were significantly underestimated. The SCF energy is sensitive to the pseudopotential as well as the exchange and correlation functional used in the DFT calculation, hence the relative energies of simulated NiO nanoparticles are dependent on ground state energy.

The frequency-dependent dielectric function is given by:

$$\varepsilon\ (\omega) = \varepsilon_1(\omega) + i\varepsilon_2(\omega) \tag{2}$$

The dielectric function is closely related to the electronic band structure and it fully describes the optical properties of any homogeneous medium at all photon energies. The imaginary part of the complex dielectric function is obtained from the momentum matrix elements between the occupied and unoccupied electronic states. The imaginary part is calculated using the analytical expression

$$\varepsilon_2(\omega) = \frac{2e^2\pi}{\Omega\varepsilon_0}\sum_{k,v,c}\left|\left\langle\psi_k^c\middle|\hat{u}\bullet\vec{r}\middle|\left\langle\psi_k^v\right|\right|^2\delta\left(E_k^c - E_k^v - E\right) \tag{3}$$

where $\omega$ is the frequency of light, $e$ is the electronic charge, $\hat{u}$ is the vector defining the polarization of the incident electric field, $\psi_k^c$ and $\psi_k^v$ are the conduction and valence band wave functions at $k$, respectively. The real part of the dielectric function $\varepsilon_1\ (\omega)$ is derived from the imaginary part of the dielectric function $\varepsilon_2\ (\omega)$ through the Kramers–Kronig relationship. Other optical properties, such as refractive index, absorption spectrum, loss function, reflectivity, and conductivity (real part), are derived from

$$\varepsilon_1(\omega) = 1 + \frac{2}{\pi}P\int_0^\infty \frac{\omega'\varepsilon_1(\omega')d\omega'}{(\omega'^2 - \omega^2)} \tag{4}$$

### 3.4. Diffuse Reflectance Spectral (DRS) Analysis

The reflectance, transmittance and scattering of the synthesized materials were determined using diffuse reflectance spectra. Figure 6 illustrates the diffuse reflectance spectra of different NiO samples. It can be seen from Figure 6a that each sample exhibits an absorption edge and the reflectance decreases with the addition of Li. The spectra were further used to plot the optical energy gap spectra of NiO samples as indicated in Figure 6b. The band gap decreases from 3.3 eV (undoped NiO) to a minimum of 3.17 eV (NiO doped with 1% Li). The variation in the band gap cannot be explained plainly on the basis of Quantum Size Effect. This is because, as seen from the XRD data, the particle size obtained is beyond quantum size confinement.

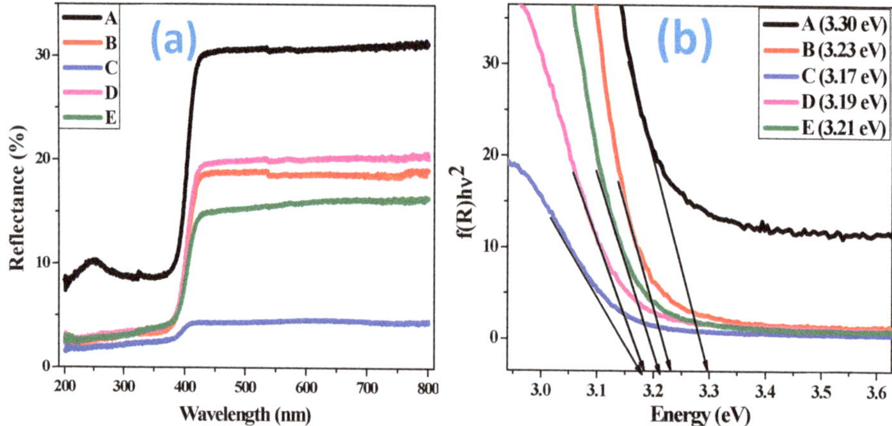


**Figure 6.** (**a**) Diffuse reflectance spectra of NiO samples. (**b**) Optical energy gap spectra of NiO samples.

With an octahedral symmetry, the divalent nickel ions possess a $3d^8$ electronic configurations and the spin-allowed d–d transitions of octahedral $Ni^{2+}$ ions are because of the single broad absorption band visible at 410 nm. As the powder sample diffuses the light in large quantities with a greater thickness, the absorption spectra becomes more complicated to interpret. In order to minimize this difficulty, DRS along with Schuster–Kubelka–Munk (SKM) relation has been used [36,37] which can be given as

$$F(R) = \frac{(1-R)^2}{2R} \tag{5}$$

where $R$ is the absolute reflectance of the sample and $F(R)$ is the Kubelka–Munk function.

Electron excitation from the valance band to conduction band gives a measure of the optical band gap which can be estimated using the relation $(F(R)\,h\upsilon)^n = A(h\upsilon\text{-}E_g)$, where $n = 2$ and ½ for directly allowed and indirectly allowed transition respectively, $A$ is the constant, and $h\upsilon$ is the photon energy [38]. Extrapolation of the linear part of the curve in Figure 6b to $(F(R)h\upsilon)^2 = 0$ leads to the direct band gap energy. The decrease in band gap can be attributed to the concentration of free carriers as well as impurity effect when NiO is doped. Normally, NiO and Li doped NiO are p-type semiconductors containing an excess of oxygen. On doping, some of the $Ni^{2+}$ are replaced with $Li^+$. This results in an increase in the concentration of $Ni^{3+}$ and holes too. According to the Moss–Burstein effect [39], this should have caused a shift in the reflectance edge towards a higher photon energy. However, in the present case, the band gap decreases due to the following two Coulomb forces: (i) the exchange and correlation energy between the holes and the electrons in the valence and conduction band, respectively; and (ii) interaction between holes and impurity ions.

Optical properties for NiO and Li doped systems were calculated using the first-principle density functional theory method. Simulated reflectivity as a function of wavelength is presented in Figure 7, along with experimental measurements. As observed experimentally, the nanoparticle depicts an absorption edge in the ultraviolet region. Optical absorption curve showed that the nanoparticles absorbance is attributed by peaks ranging from 100 to 400 nm. This suggests that the Li doped NiO particles can be stable at wavelengths below 400 nm. Furthermore, simulations showed a long wavelength activity in the visible light region.

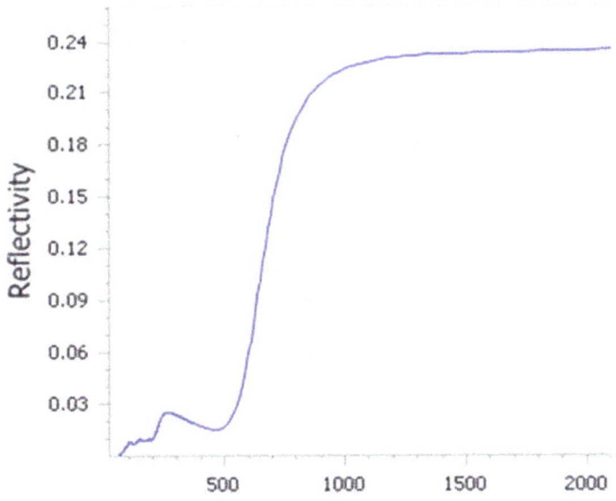


**Figure 7.** Simulated reflectivity as a function of wavelength.

### 3.5. Photoluminescence Studies (PL)

The photoluminescence excitation spectra (Figure 8) of the prepared samples at emission wavelength of 410 nm exhibit a prominent broad band at 308 nm along with an intense peak at 371 nm, corresponding, respectively, to spin-allowed $3T_1g(F) \leftarrow 3A_2g$and$3T_1g(P) \leftarrow 3A_2g$ transitions of $Ni^{2+}$ ions [40]. The intensity and position of these bands are characteristic of octahedral $Ni^{2+}$.

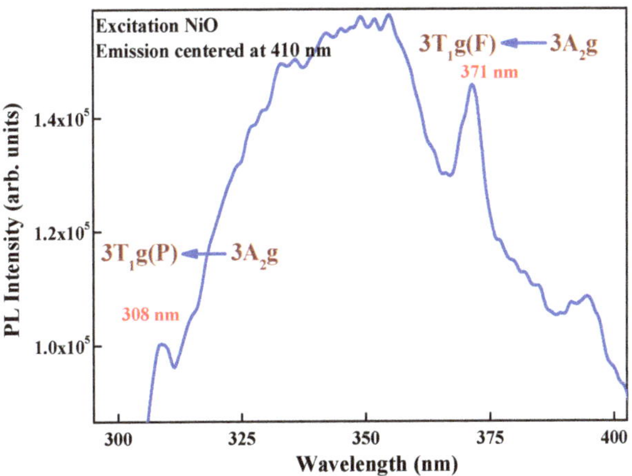


**Figure 8.** Excitation spectrum of NiO emission wavelength monitored at 410 nm.

The PL emission spectra (Figure 9) of the samples, upon excitation at 308 nm, shows two PL peaks that are obvious at 345 and 466 nm, corresponding to ultraviolet emission (340–400 nm) and blue emission (450–495 nm), respectively. Similar observations have been made in the literature citing PL studies on NiO nanoparticles [41,42].The origin of photoluminescence peaks can be attributed to electronic transitions involving $3d^8$ electrons of the $Ni^{2+}$ ions [43].The presence of $Li^+$ resulted in the formation of several shoulder peaks: at 402 and 422 nm (violet emission), 452 nm (blue emission) and 508 nm (green emission).The violet emission peaks are probably due to the transition of trapped electrons at interstitial Ni to the valence band. The blue emissions are due to the recombination of electrons from the $Ni^{2+}$ vacancy to the holes in the valence band [41]. The cause of the green emission peak is not yet clear, as some authors cite it to be due to increases in Ni vacancies [44] whereas others relate it to the oxygen vacancies [45]. The addition of lithium influences the PL spectra profoundly; samples D and E have a lower emission intensity, whereas sample C has a higher emission intensity, than pristine NiO. The UV and visible emission peak intensity is dependent on radiative recombination. At a lower concentration of dopant, the radiative recombination process is higher, emitting more energy and thereby intensifying the peak. A small amount of dopant results in the replacement of $Ni^{2+}$ with $Li^+$, generating one hole in the valence band to maintain charge neutrality [46]. However, as the dopant concentration increases, it induces higher defects. This probably leads to non-radiative recombination, which subsequently reduces the peak intensity [47]. Interestingly, the sample B (NiO–surfactant) exhibits a high-intensity broad peak from 400–550 nm with maxima at 440 nm. A similar result has been reported by Wang and his group for NiO synthesized using dodecylamine as surfactant. The authors have attributed the broad nature of PL emission to the multilayer structure formed by layered NiO–surfactant superlattices. This kind of structure is expected to influence the chemical and physical properties of NiO to a large extent [48]. The oxygen vacancies may interact with interfacial capping surfactants, forming a series of metastable energy levels within the band gap. The long lifetime and dipole allowed for by transitions in these metastable energy levels induces the interfacial effect between NiO and the surfactant. Such unusual room temperature photoluminescences have also been previously observed by Zou and his group for nanoparticles coated with stearic acid [49] and by Bai and co-workers for mesolamellar $TiO_2$ structures [50]. The intensity emission peak indicates enhanced photoluminescence intensity with high charge transfer resistance and, hence, a decrease in the electrochemical behavior further confirmed by CV and EIS analysis.

### 3.6. *Cyclic VoltammetryAnalysis*

The cyclic voltammetric studies of the synthesized samples are shown in Figure 10. The CV curve has definite symmetric anodic and cathodic peaks, indicating the good reversibility of the redox reactions. However, by increasing the scan rates, no significant change was observed within the material. As evidenced by CV studies, the electrochemical process of NiO electrodes is limited by the proton diffusion through the lattice [51–53]. By taking into account the difference between the oxidation potential ($E_O$) and the reduction potential ($E_R$) at a scan rate of 10 mV/s, the reversibility of the electrode reaction was measured. The reversibility increases as the $E_O - E_R$ value decreases. From Table 2, it is evident that the reversibility of the electrode reaction was maximum for NiO sample prepared with 1% Li dopant. The electrode reactions in NiO proceed according to the following reaction:

$$\mathrm{NiO} \underset{\text{Reduction}}{\overset{\text{Oxidation}}{\rightleftharpoons}} \mathrm{Ni}^{2+} + \mathrm{O}_2^{2-} \tag{6}$$

The anodic and cathodic peaks indicate the oxidation of $Ni^0$ into $Ni^{2+}$ and reduction of $Ni^{2+}$ into $Ni^0$, respectively. The quasi-reversible electron transfer process seen in the CV curve indicates the measured capacitance based on the redox mechanism [54]. The peak heights or the area of the CV curve for doped samples are larger than pristine samples, denoting a high amount of stored charge. This could be related to the smaller size of the doped samples resulting in a higher surface area. However, NiO with 2% Li exhibits a smaller CV curve when compared to the undoped samples. The reason for this deviation is unclear.

The quantity of current generated by electrode C is comparatively higher, and is least for electrode B. This suggests that 1% Li plays a crucial role during lattice formation which is manifested by generation of higher current when compared to the rest of the doped oxides. On the other hand, the presence of PEG as a surfactant during the synthesis has a great impact on the surface morphology. This is explicit by a higher PL intensity, as well as a decrease in the electrode reversibility. Probably, the surfactant enclosing NiO diminishes the development of the current.

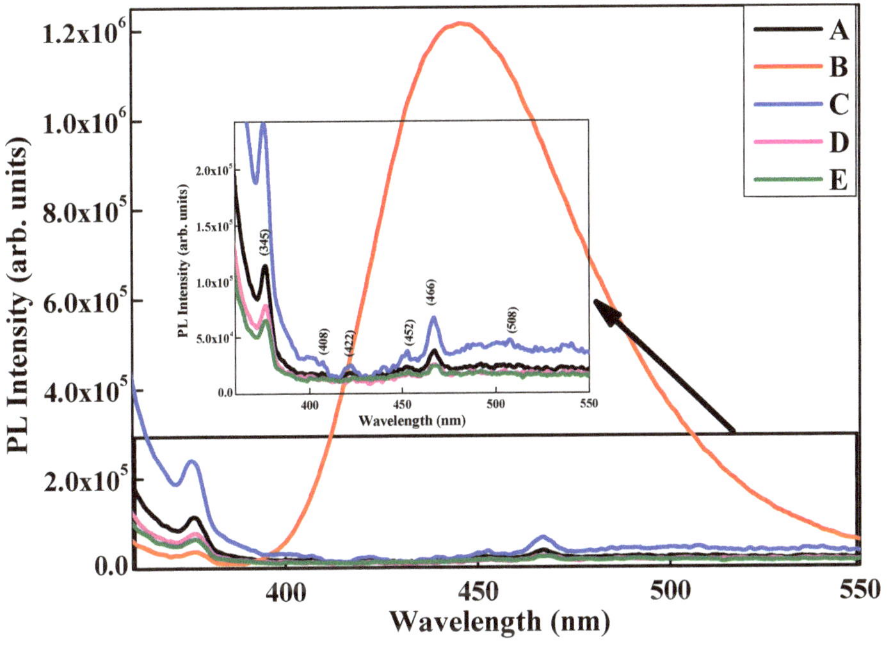


**Figure 9.** Emission spectra of different NiO samples excited at 308 nm.

**Table 2.** Oxidation potential ($E_O$), reduction potential ($E_R$) and the difference between $E_O$ and $E_R$ of different NiO electrodes.

| Name of the Electrode | $E_O$ | $E_R$ | $E_O - E_R$ |
|---|---|---|---|
| A | 0.419 | 0.240 | 0.179 |
| B | 0.428 | 0.201 | 0.227 |
| C | 0.392 | 0.238 | 0.154 |
| D | 0.419 | 0.253 | 0.166 |
| E | 0.428 | 0.254 | 0.174 |

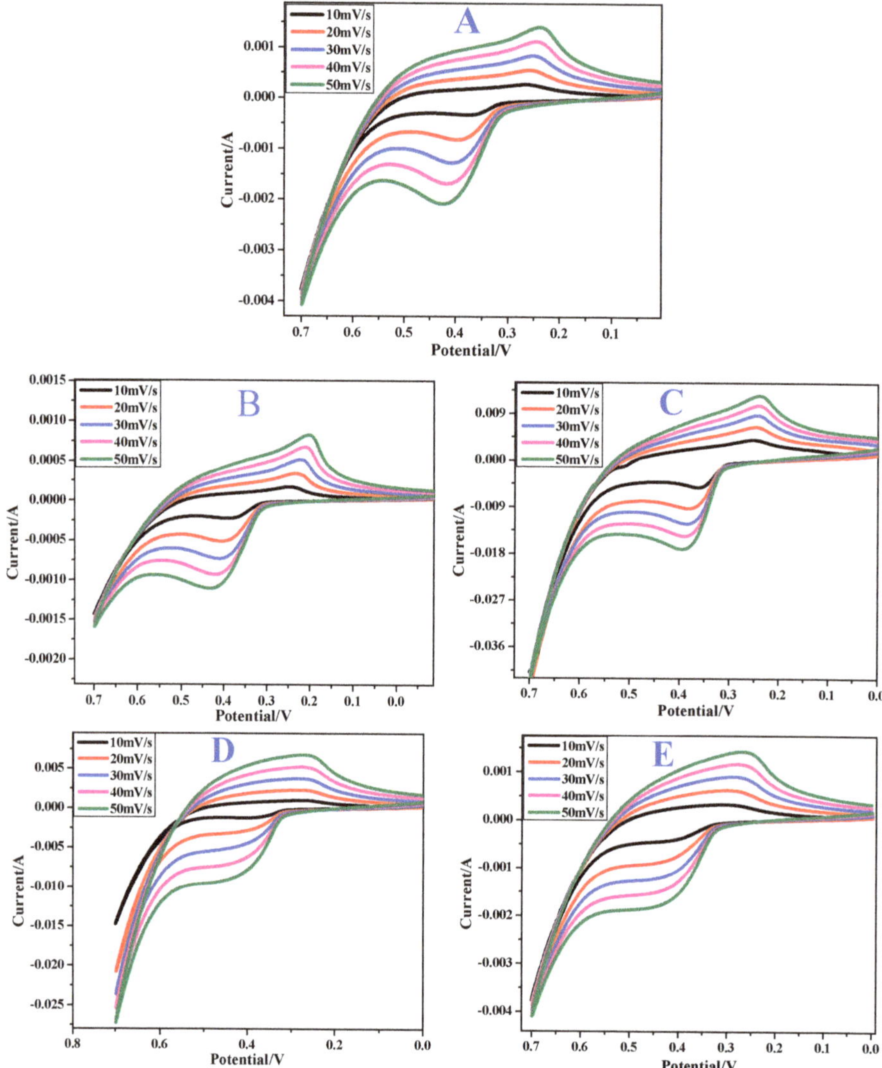


**Figure 10.** Cyclic voltammogram of (**A**) pristine NiO, (**B**) pristine NiO with PEG, (**C**) NiO with PEG and 1 wt% Li, (**D**) NiO with PEG and 2 wt% Li, (**E**) NiO with PEG and 5 wt% Li at various scan rates.

### *3.7. Impedance Studies*

The electrochemical impedance measurements were carried out in the frequency range of 1 Hz to 1 MHz at 5 mV steady state amplitude. The corresponding Nyquist plots are shown in Figure 11. The plots suggest a larger impedance for electrode A (semicircle with a bigger diameter) while the impedance of the electrode C was found to be smaller (Table 3), as is evident by the shifting of the imaginary line towards Y-axis. Consequently, electrode C manifests higher discharge rates and capacitance. The increase in the capacitance can also be attributed to the combined effect of the electric double-layer capacitance on the high surface area of AC and pseudocapacitance via the intercalation/extraction of Li ions in NiO lattice.

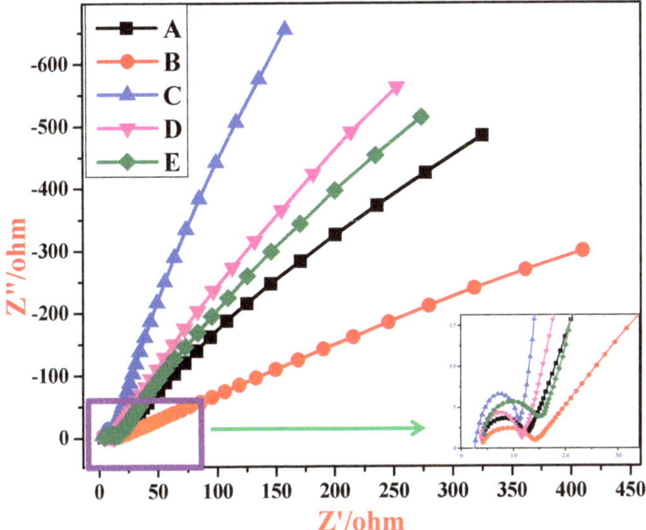


**Figure 11.** Nyquist plots of samples A, B, C, D and E.

**Table 3.** The EIS fitted circuit parameters of $R_{Ct}$ and $C_{dl}$ values.

| Name of the Electrode | $R_{Ct}$ (Ω) | $C_{dl}$ (F) |
|---|---|---|
| A | $9.146 \times 10^{-3}$ | $1.147 \times 10^{-4}$ |
| B | $3.549 \times 10^{-2}$ | $1.404 \times 10^{-5}$ |
| C | $5.592 \times 10^{-8}$ | $1.72 \times 10^{-3}$ |
| D | $1.277 \times 10^{-4}$ | $1.239 \times 10^{-5}$ |
| E | $8.091 \times 10^{-4}$ | $1.201 \times 10^{-7}$ |

Since the Nyquist plots reveal the presence of depressed semicircles with a centre below the real axis at higher frequencies, it becomes necessary to use a constant phase element ($Q_1$) to fit the data into an equivalent circuit. The impedance of $Q_1$ can be described [55,56] as

$$Z_{CPE} = \frac{1}{Y(j\omega)^n} \tag{7}$$

where $\omega$ is the angular frequency in rad s$^{-1}$, $Y$ and $n$ are adjustable parameters of constant phase element ($Q_1$). For double layer capacitance, the value of $n = 1$, for resistance and Warburg diffusion $n = 0$ and $n = 0.5$, respectively.

The equivalent circuit for the Nyquist plots of impedance measurements of NiO electrodes A–E is shown in Figure 12. In the given circuit, the high frequency region corresponds to solution resistance ($R_s$) at the electrode–electrolyte interface. The semicircles are attributed to an interfacial charge transfer resistance ($R_{ct}$) or polarization resistance ($R_p$) to which the double-layer capacitance (C) is connected in parallel. The low frequency region straight line is represented by the Warburg element (W) in series with $R_{ct}$. Warburg element is an estimation of the redox reactions occurring in the system, i.e., the diffusion of electrons from the working electrode and deposition of nickel ions from the electrolyte into the pores on the electrode surface [57], the electrolytic diffusion of ions takes place during the transition from the high-frequency semicircle to the mid-frequency point [58]. The constant phase element ($Q_1$) lies parallel to the charge-transfer resistance ($R_{ct}$), as does the low frequency capacitance ($Q_2$) to the leakage resistance ($R_l$).

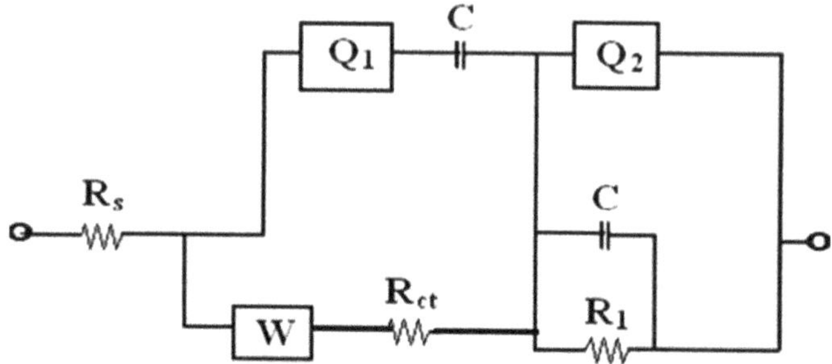


**Figure 12.** Equivalent circuit for Nyquist plot of samples A, B, C, D and E.

EIS-fitted circuit parameters are tabulated in Table 3. The data were obtained by fitting the experimental data as per the equivalent circuit. A decrease in $R_{ct}$ and an increase in $C_{dl}$ indicates an enhancement in the electrochemical activity of the electrode. It is evident from Table 3 that the electrochemical activity of the electrode C (1% Li dopant) was higher, which may be attributed to the Li grains' effectiveness in current collection and further improves the charge transfer process on interface between electrolyte and electrode. However, this effect subsides with increasing Li concentration. On the other hand, due to the surfactant effect there is an increase in the $R_{ct}$ value and a decrease in $C_{dl}$ value. This is consistent with the PL emission results.

## 4. Conclusions

In summary, undoped and Li doped NiO nanoparticles have been successfully prepared by microwave technique. The nanostructures were found to be crystalline in nature, with a cubic structure. The crystallite size decreased after the introduction of an Li dopant as revealed from X-ray diffraction studies. A decrease in band gap and an ultraviolet-blue emission along with small amount of green emission suggests that the photoluminescence of NiO nanomaterials can be tuned by doping. The optical results were further confirmed by computational modelling. From CV measurements, it was observed that the reversibility of NiO sample was maximum with 1% Li dopant. The same electrode also had enhanced electrochemical property with the charge transfer resistance reducing to $5.592 \times 10^{-8}$ Ω. Through the present work, the optical and electrochemical behavior of Li doped NiO have been successfully demonstrated. We believe that these nanostructures can be applied to nanoscale electrochromic and electro-optical devices for various consumer and industrial applications.

**Author Contributions:** Conceptualization, A.S.B.; methodology, R.R.; software, R.R.M.; validation, A.S.B., R.R., C.R.R. and S.C.P.; formal analysis, A.S.B. and M.S.S.; investigation, R.R.; resources, C.R.R., S.C.P., M.S.S. and G.F.B.L.eS.; data curation, A.S.B.; writing—original draft preparation, A.S.B. and C.R.R.; writing—review and editing, M.S.S., G.F.B.L.eS.; visualization, R.R.; supervision, A.S.B. and M.S.S.; project administration, A.S.B.; funding acquisition, A.S.B. All authors have read and agreed to the published version of the manuscript.

**Funding:** Aarti S. Bhatt thanks the VGST, Govt. of Karnataka, India, (VGST/SMYSR/GRD-437/2014-15) for extending financial helps to carry out this research work. All the authors thank the University of Sao Paulo, Brazil for funding the APC.

**Conflicts of Interest:** The Authors declares that they have no Conflict of Interest.

## References

1. Yang, H.; Tao, Q.; Zhang, X.; Tang, A.; Ouyang, J. Solid-state synthesis and electrochemical property of $SnO_2$-NiO nanomaterials. *J. Alloys Compd.* **2008**, *459*, 98. [CrossRef]
2. Fujii, E.; Tomozawa, A.; Torii, H.; Takayama, R. Preferred orientations of NiO films prepared by plasma enhanced metal organic chemical vapor deposition. *J. Appl. Phys.* **1996**, *35*, 328. [CrossRef]

3. Sato, H.; Minami, T.; Takata, S.; Yamada, T. Transparent conducting p-type NiO thin films prepared by magnetron sputtering. *Thin Solid Films* **1993**, *236*, 27. [CrossRef]
4. Kitao, M.; Izawa, K.; Urabe, K.; Komatsu, T.; Kuwano, S.; Yamada, S. Preparation and electrochromic properties of RF-sputtered NiOx films prepared in Ar/O2/H2 atmosphere. *J. Appl. Phys.* **1994**, *33*, 6656. [CrossRef]
5. Kumagai, H.; Matsumoto, M.; Toyoda, K.; Obara, M. Preparation and characteristics of nickel oxide thin film by controlled growth with sequential surface chemical reactions. *J. Mater. Sci. Lett.* **1996**, *15*, 1081. [CrossRef]
6. Ellmer, K.; Mientus, R. Carrier transport in polycrystalline transparent conductive oxides: A comparative study of zinc oxide and indium oxide. *Thin Solid Films* **2008**, *516*, 4620. [CrossRef]
7. Xiang, H.J.; Yang, J.; Hou, J.G.; Zhu, Q. Piezoelectricity in ZnO nanowires: A first principles study. *Appl. Phys. Lett.* **2006**, *89*, 223111. [CrossRef]
8. Yi, J.B.; Lim, C.C.; Xing, G.Z.; Fan, H.M.; Van, L.H.; Huang, S.L.; Yang, K.S.; Huang, X.L.; Qin, X.B.; Wang, B.Y.; et al. Ferromagnetism in dilute magnetic semi-conductors through defect engineering: Li-doped ZnO. *Phys. Rev. Lett.* **2010**, *104*, 137201. [CrossRef]
9. Guo, W.; Hui, K.N.; Hui, K.S. High conductivity nickel oxide thin films by a facile sol-gel method. *Mater. Lett.* **2013**, *92*, 291. [CrossRef]
10. Jang, W.L.; Lu, Y.M.; Hwang, W.S.; Chen, W.C. Electrical properties of Li doped NiO films. *J. Eur. Ceram. Soc.* **2010**, *30*, 503–508. [CrossRef]
11. Moulki, H.; Park, D.; Min, B.K.; Kwon, H.; Hwang, S.J.; HoChoy, J.; Toupance, T.; Campet, G.; Rougier, A. Improved electrochromic performances of NiO based thin films by lithium addition: From single layers to devices. *Electrochim. Acta* **2012**, *74*, 46. [CrossRef]
12. Lakshmi Devi, L.; Jayasankar, C.K. Spectroscopic investigations on high efficiency deep red emitting $Ca_2SiO_4$:$Eu^{3+}$ phosphors synthesized from agricultural waste. *Ceram. Int.* **2018**, *44*, 14063. [CrossRef]
13. Furuya, S.; Nakano, H.; Yokoyoma, N.; Banno, H.; Fukuda, K. Enhancement of photoluminescence intensity and structural change by doping $P^{5+}$ ion for $Ca_{2-x/2}(Si_{1-x}P_x)O_4$:$Eu^{2+}$ green phosphor. *J. Alloys Compd.* **2016**, *658*, 147. [CrossRef]
14. Nakano, H.; Kamimoto, K.; Yokoyama, N.; Fukuda, K. The effect of heat treatment on the emission color of P-doped $Ca_2SiO_4$ phosphor. *Materials* **2017**, *10*, 1000. [CrossRef]
15. Rojas-Hernandez, R.E.; Rubio-Marcos, F.; Serrrano, A.; Salas, E.; Hussainova, I.; Fernandez, J.F. Towards blue long lasting luminescence of Eu/Nd-doped calcium aluminate nanostructured platelets via the molten salt route. *Nanomaterials* **2019**, *9*, 1473. [CrossRef]
16. Shionoya, S.; Yen, W.M. *Phosphor Handbook*; CRC: Boca Raton, FL, USA, 1999; p. 579.
17. Leskel, M. Rare earths in electroluminescent and field emission display phosphors. *J. Alloys Compd.* **1998**, *275*, 702. [CrossRef]
18. Ortiz, A.; Falcony, C.; Hernandez, J.; Garcia, A.M.; Alonso, J.C. Photoluminescence characteristics of lithium doped zinc oxide films deposited by spray pyrolysis. *Thin Solid Films* **1997**, *293*, 103. [CrossRef]
19. Ohashi, N.; Nakata, T.; Sekiguchi, T.; Hosono, H.; Mizuguchi, M.; Tsurumi, T.; Tanaka, J.; Haneda, H. Yellow emission from zinc oxide giving an electron spin resonance signal at g = 1.96. *Jpn. J. Appl. Phys.* **1999**, *38*, L113. [CrossRef]
20. Fujihara, S.; Ogawa, Y.; Kasai, A. Tunable visible photoluminescence from ZnO thin films through Mg-doping and annealing. *Chem. Mater.* **2004**, *16*, 2965. [CrossRef]
21. Mordkovich, V.Z.; Hayashi, H.; Haemori, M.; Fukumura, T.; Kawasaki, M. Discovery and optimization of new ZnO-based phosphors using a combinatorial method. *Adver. Funct. Mater.* **2003**, *13*, 519. [CrossRef]
22. Lamrani, M.A.; Addou, M.; Sofiani, Z.; Sahraoui, B.; Ebothe, J.; Hichou, A.E.; Fellahi, N.; Bernede, J.C.; Dounia, R. Cathodoluminescent and nonlinear optical properties of undoped and erbium doped nanostructured ZnO films deposited by spray pyrolysis. *Opt. Commun.* **2007**, *277*, 196. [CrossRef]
23. Jeevanandam, P.; Pulimi, V.R.R. Synthesis of nanocrystalline NiO by sol-gel and homogeneous precipitation methods. *Indian J. Chem.* **2012**, *51A*, 586.
24. Benedict, S.; Singh, M.; Nail, T.R.R.; Shivashankar, S.A.; Bhat, N. Microwave synthesized NiO as a highly sensitive and selective room temperature $NO_2$ sensor. *ECS J. Solid State Sci. Technol.* **2018**, *7*, Q3143. [CrossRef]
25. Raj, R.A.; Alsalhi, M.S.; Devanesan, S. Microwave-assisted synthesis of nickel oxide nanoparticles using coriandrum sativum leaf extract and their structural-magnetic catalytic properties. *Materials* **2017**, *10*, 460. [CrossRef]

26. Parada, C.A.; Moran, E. Microwave-assisted synthesis and magnetic study of nanosized Ni/NiO materials. *Chem. Mater.* **2006**, *18*, 2719. [CrossRef]
27. Li, X.; Tan, J.F.; Hu, Y.E.; Huang, X.T. Microwave-assisted synthesis of Fe-doped NiO nanofoams assembled by porous nanosheets for fast response and recovery gas sensors. *Mater. Res. Express* **2017**, *4*, 045015. [CrossRef]
28. Segall, M.D.; Lindan, J.; Peakard, C.J.; Hasnip, P.J.; Clark, S.J.; Payne, M.C. First principles simulation: Ideas, illustration and the CASTEP code. *J. Phys. Cond. Matter* **2002**, *14*, 2717. [CrossRef]
29. Perdew, P.; Wang, Y. Accurate and simple density functional for the electronic exchange energy: Generalized gradient approximation. *Phys. Rev. B* **1986**, *33*, 8800. [CrossRef]
30. Shanon, R.D. Revised effective ionic radii and systematic studies of interatomic distances in halides and chalcogenides. *Acta Cryst. A* **1976**, *32*, 751. [CrossRef]
31. Chia, C.E.; Cheng-Fu, Y. Investigation of the properties of nanostructured Li-doped NiO films using the modified spray pyrolysis method. *Nanoscale Res. Lett.* **2013**, *8*, 33. [CrossRef]
32. Matsubara, J.; Huang, S.; Iwamoto, M.; Pan, W. Enhanced conductivity and gating effect of p-type Li-doped NiO nanowires. *Nanoscale* **2014**, *6*, 688. [CrossRef] [PubMed]
33. Bartel, L.C.; Morosin, B. Exchange striction in NiO. *Phys. Rev. B* **1971**, *3*, 1039. [CrossRef]
34. Tasker, P.W.; Duffy, D.M. The structure and properties of stepped surfaces of MgO and NiO. *Surf. Sci.* **1984**, *137*, 91. [CrossRef]
35. Xiang, J.; Xiang, B.; Cui, X. NiO nanoparticle surface energy studies using first principles calculations. *New J. Chem.* **2018**, *42*, 10791. [CrossRef]
36. Tripathi, N.; Shyama, R. Facile synthesis of ZnO nanostructures and investigation of structural and optical properties. *Mater. Char.* **2013**, *86*, 263. [CrossRef]
37. Pratapkumar, C.; Prashantha, S.C.; Nagabhushana, H.; Anilkumar, M.R.; Ravikumar, C.R.; Nagaswarupa, H.P.; Jnaneshwara, D.M. White light emitting magnesium aluminate nanophosphor: Near ultra violet excited photoluminescence, photometric characteristics and its UV photocatalytic activity. *J. Alloys Compd.* **2017**, *728*, 1124. [CrossRef]
38. Lopez, R.; Gomez, R. Band-gap energy estimation from diffuse reflectance measurements on sol-gel and commercial $TiO_2$: A comparative study. *J. Sol–Gel Sci. Technol.* **2012**, *61*, 1. [CrossRef]
39. Saha, B.; Das, S.; Chattopadhyay, K.K. Electrical and optical properties of Al doped cadmium oxide thin films deposited by radio frequency magnetron sputtering. *Solar Energy Mater. Sol. Cel.* **2007**, *91*, 1692. [CrossRef]
40. Ikeda, K.; Vedanand, S. Optical spectrum of synthetic theophrastite $Ni(OH)_2$. *Neues Jahrb. Mineral. Monatsh.* **1999**, *1*, 21.
41. Gandhi, A.C.; Wu, S.Y. Strong deep-level-emission photoluminescence in NiO nanoparticles. *Nanomaterials* **2017**, *7*, 231. [CrossRef]
42. Musevi, S.J.; Aslani, A.; Motahari, H.; Salimi, H. Offer a novel method for size appraise of NiO nanoparticles by PL analysis: Synthesis by sonochemical method. *J. Saudi Chem. Soc.* **2016**, *20*, 245. [CrossRef]
43. Anandan, K.; Rajendran, V. Morphological and size effects of NiO nanoparticles via solvothermal process and their optical properties. *Mater. Sci. Semicond. Process* **2011**, *14*, 43. [CrossRef]
44. Jiang, D.Y.; Qin, J.M.; Wang, X.; Gao, S.; Liang, Q.C.; Zhao, J.X. Optical properties of nickel oxide thin films fabricated by electron beam evaporation. *Vacuum* **2012**, *86*, 1083. [CrossRef]
45. Madhu, G.; Biju, V. Effect of $Ni^{2+}$ and $O^{2-}$ vacancies on the electrical and optical properties of nanostructured nickel oxide synthesized through a facile chemical route. *Physica E* **2014**, *60*, 200. [CrossRef]
46. Chen, H.R.; Harding, J.H. Nature of the hole states in Li-doped NiO. *Phys. Rev. B* **2012**, *85*, 115127. [CrossRef]
47. Jain, S.; Khare, P.S.; Pandey, D.K. Synthesis and analysis of photoluminescence properties of In and Sn doped nickel oxide powders. *Optik* **2016**, *127*, 6763. [CrossRef]
48. Wang, Y.; Ma, C.; Sun, X.; Li, H. Preparation and photoluminescence properties of organic-inorganic nanocomposite with a mesolamellar nickel oxide. *Microporous Mesoporous Mater.* **2004**, *71*, 99. [CrossRef]
49. Zou, B.; Xiao, L.; Li, T.J.; Zhao, J.; Lai, Z.; Gu, S. Absorption redshift in $TiO_2$ ultrafine particles with surficial dipole layer. *Appl. Phys. Lett.* **1991**, *59*, 1826. [CrossRef]
50. Bai, N.; Li, S.G.; Chen, H.Y.; Peng, W.Q. Preparation, characterization and photoluminescence properties of mesolamellar titanium dioxide films. *Mater. Chem.* **2001**, *11*, 3099. [CrossRef]
51. Bernard, M.C.; Cortes, R.; Keddam, M.; Takenouti, H.; Bernard, P.; Senyarich, S. Structural defects and electrochemical reactivity of $\beta$-$Ni(OH)_2$. *J. Power Sources* **1996**, *63*, 247. [CrossRef]

52. Baraldi, P.; Davolio, G. An electrochemical and spectral study of the nickel oxide electrode. *Mater. Chem. Phys.* **1989**, *21*, 143. [CrossRef]
53. Ravikumar, C.R.; Anil Kumar, M.R.; Nagaswarupa, H.P.; Prashantha, S.C.; Bhatt, A.S.; Santosh, M.S.; Kuznetsov, D. CuO embedded β-$Ni(OH)_2$ nanocomposite as advanced electrode materials for supercapacitors. *J. Alloys Compd.* **2018**, *736*, 332–339. [CrossRef]
54. Bard, A.J.; Faulkner, L.R. *Electrochemical Methods: Fundamentals and App*, 2nd ed.; John Wiley and Sons: Hoboken, NJ, USA, 2000; p. 368.
55. Brug, G.J.; Eeden, V.D.; Sluyters-Rehobach, M.; Sluyters, J.H. The analysis of electrode impedances complicated by the presence of a constant phase element. *J. Electroanal. Chem.* **1984**, *176*, 275. [CrossRef]
56. Masarapu, C.; Zeng, H.F.; Hung, K.H.; Wei, B. Effect of temperature on the capacitance of carbon nanotube capacitors. *ACS Nano* **2009**, *3*, 2199. [CrossRef]
57. Yen Ho, M.; Khiew, S.P.; Isa, D.; Chiu, W.S. Electrochemical studies on nanometal oxide-activated carbon composite electrodes for aqueous supercapacitors. *Funct. Mater. Lett.* **2014**, *7*, 1440012. [CrossRef]
58. Avinash, B.; Ravikumar, C.R.; Anil Kumar, M.R.; Nagaswarupa, H.P.; Santosh, M.S.; Bhatt, A.S.; Kuznetsov, D. Nano CuO: Electrochemical sensor for the determination of paracetamol and D-glucose. *J. Phys. Chem. Solids* **2019**, *134*, 193. [CrossRef]

*Article*

# A Study of Methylcellulose Based Polymer Electrolyte Impregnated with Potassium Ion Conducting Carrier: Impedance, EEC Modeling, FTIR, Dielectric, and Device Characteristics

**Muaffaq M. Nofal [1], Jihad M. Hadi [2], Shujahadeen B. Aziz [3,4,*], Mohamad A. Brza [3], Ahmad S. F. M. Asnawi [5], Elham M. A. Dannoun [6], Aziz M. Abdullah [7] and Mohd F. Z. Kadir [8]**

1 Department of Mathematics and General Sciences, Prince Sultan University, P.O. Box 66833, Riyadh 11586, Saudi Arabia; muaffaqnofal69@gmail.com
2 Department of Medical Laboratory of Science, College of Health Sciences, University of Human Development, Sulaimaniyah 46001, Iraq; jihad.chemist@gmail.com
3 Hameed Majid Advanced Polymeric Materials Research Laboratory, Physics Department, College of Science, University of Sulaimani, Qlyasan Street, Sulaimaniyah 46001, Iraq; mohamad.brza@gmail.com
4 Department of Civil Engineering, College of Engineering, Komar University of Science and Technology, Sulaimaniyah 46001, Iraq
5 Chemical Engineering Section, Malaysian Institute of Chemical & Bioengineering Technology (UniKL MICET), University Kuala Lumpur, Alor Gajah 78000, Malaysia; asyafiq.asnawi@s.unikl.edu.my
6 General Science Department, Woman Campus, Prince Sultan University, P.O. Box 66833, Riyadh 11586, Saudi Arabia; elhamdannoun1977@gmail.com
7 Department of Applied Physics, College of Medical and Applied Sciences, Charmo University, Peshawa Street, Chamchamal 46023, Iraq; aziz.abdullah@charmouniversity.org
8 Centre for Foundation Studies in Science, University of Malaya, Kuala Lumpur 50603, Malaysia; mfzkadir@um.edu.my
* Correspondence: shujahadeenaziz@gmail.com



**Citation:** Nofal, M.M.; Hadi, J.M.; Aziz, S.B.; Brza, M.A.; Asnawi, A.S.F.M.; Dannoun, E.M.A.; Abdullah, A.M.; Kadir, M.F.Z. A Study of Methylcellulose Based Polymer Electrolyte Impregnated with Potassium Ion Conducting Carrier: Impedance, EEC Modeling, FTIR, Dielectric, and Device Characteristics. *Materials* **2021**, *14*, 4859. https://doi.org/10.3390/ma14174859

Academic Editors: Diogo Miguel Franco dos Santos and Biljana Šljukić

Received: 2 June 2021
Accepted: 20 August 2021
Published: 26 August 2021



**Abstract:** In this research, a biopolymer-based electrolyte system involving methylcellulose (MC) as a host polymeric material and potassium iodide (KI) salt as the ionic source was prepared by solution cast technique. The electrolyte with the highest conductivity was used for device application of electrochemical double-layer capacitor (EDLC) with high specific capacitance. The electrical, structural, and electrochemical characteristics of the electrolyte systems were investigated using various techniques. According to electrochemical impedance spectroscopy (EIS), the bulk resistance ($R_b$) decreased from $3.3 \times 10^5$ to $8 \times 10^2$ $\Omega$ with the increase of salt concentration from 10 wt % to 40 wt % and the ionic conductivity was found to be $1.93 \times 10^{-5}$ S/cm. The dielectric analysis further verified the conductivity trends. Low-frequency regions showed high dielectric constant, $\varepsilon'$ and loss, $\varepsilon''$ values. The polymer-salt complexation between (MC) and (KI) was shown through a Fourier transformed infrared spectroscopy (FTIR) studies. The analysis of transference number measurement (TNM) supported ions were predominantly responsible for the transport process in the MC-KI electrolyte. The highest conducting sample was observed to be electrochemically constant as the potential was swept linearly up to 1.8 V using linear sweep voltammetry (LSV). The cyclic voltammetry (CV) profile reveals the absence of a redox peak, indicating the presence of a charge double-layer between the surface of activated carbon electrodes and electrolytes. The maximum specific capacitance, $C_s$ value was obtained as 118.4 F/g at the sweep rate of 10 mV/s.

**Keywords:** MC polymer electrolyte; impedance study; ion transport; ftir analysis; TNM; LSV; CV analyses

## 1. Introduction

Besides improving energy and power efficiency, one of the remaining challenges in the development of energy storage systems, including smart grids, portable electronic devices, and hybrid vehicles, is to minimize manufacturing costs and reduce environmental pollution [1]. A more recent emphasis has been focused on solid polymer electrolytes (SPEs) as an alternative conventional organic sol–gel electrolyte. Dimensional stability, durability, a comparatively wide potential window above 1.5 V, and eco-friendliness are all properties of these materials [2]. In the technology area, natural polymers for fabricating SPEs have gained interest for application in electrochemical devices such as electrical double-layer capacitors (EDLCs) and proton batteries. Due to their exceptional chemical and mechanical performances, many studies have shown that natural SPEs exhibit a good potential for device applications [3,4]. Natural polymers are defined as materials that extensively happen in nature or are obtained from animals or plants. Natural polymers are vital to way of life as our human forms are based on them. Some of the examples of natural polymers are nucleic acid and proteins that happen in human body, natural rubber, silk, and methylcellulose (MC). MC is known to be competitively marketed and is environmentally safe. It has suitable film-forming characteristics with good mechanical and electrical properties. Through dative bonds, cations can interact with oxygen atoms of MC. As a consequence, MC comprises functional groups, such as alcohol (R-OH), ether (R-O-R), and ester (RCOOR) groups which are promising as an ion conduction mechanism due to their single pair of electrons. MC is also considered an amorphous polymer with its comparatively high glass transition temperature [5,6].

Supercapacitors consist of two porous electrodes separated by an ionically conducting electrolyte. The electrodes could be made of substances including polymers, carbon and metal oxides. Supercapacitors can be a favorable energy conversion device for a wide range of applications, where significant amounts of energy must be stored or released in a short period. A supercapacitor is classified into three major types, namely pseudo-capacitors, EDLCs, and hybrid capacitors. Pseudo-capacitors undergo a fast Faradaic mechanism [7], some examples of which include under potential deposition, intercalation, and reduction-oxidation reactions using metal oxide-based electrodes or electroactive conducting polymer. However, EDLCs do not involve any Faradaic mechanisms. EDLCs only require the accumulation of ions induced by the adsorption of charge carrier at the electrode/electrolyte interfaces. Owing to the storage process, EDLC is the non-Faradaic mechanism [8]. The main features of EDLCs, such as reliability, high energy capacity, reversibility, and safety improvements have drawn considerable interest, and making it a strong choice for various applications [9].

Activated carbon electrodes play a crucial role in the fabricating of EDLC due to their good chemical and physical properties such as low cost and easy availability, and high conductivity above $10^{-4}$ S/cm, which can be manufactured from a diversity of precursors [10,11]. As a result, coal is the most common supply of activated carbon production due to its availability, high content of carbons from 60% to 80%, and cost-effectiveness [12,13].

T.-Y. Chen et al. [14] electrodeposited NiSe nanoparticles on a carbon nanotube (CNT) forest to prepare a porous and intertwined network (denoted as CNT@NiSe/stainless steel (SS)). They then used the CNT@NiSe/SS as a free-standing and multifunctional electrode for supercapacitor (SC) application. The CNT@NiSe/SS composite electrode showed excellent capacity retention of 85%, and higher specific capacity of 126 mA h $g^{-1}$ (1007 F $g^{-1}$) in comparison with individual CNTs and NiSe. Lien et al. [15] developed a co-solvent-in-deep eutectic solvent (DES) system by mixing acetonitrile and water with a typical DES electrolyte composed of lithium perchlorate and acetamide. They have also used hydrogel composed of reduced graphene oxide (rGO) and 1T(trigonal)-MoS2 as the electrode materials for SC application. The authors fabricated high voltage symmetric supercapacitors using hydrogel and hybrid DES as the electrode and electrolyte materials, respectively. The SC at an operating voltage of 2.3 V achieved the maximum energy density of 31.2 Wh/kg at a power density of 1164 W/kg. The fabricated SC also showed 91% capacitance retention after 20,000 cycles. Hsiang et al. [16] presented rationally materials design of an optimum $NiCo_2S_4$ nanoparticle in a rGO matrix as a $NiCo_2S_4$/rGO nanocomposite. The authors reported the enhancements in the materials technology, showing the $NiCo_2S_4$/rGO nanocomposite electrode material with a very good specific capacitance of 963–700 F/g at 1–15 A/g, long cycle life of 3000 cycles, and high capacitance retention of 70%.

Adding inorganic salt to a polymer provides ion mobility and the polymer host chain plays a crucial role in the ion transport mechanism of the polymer electrolytes. Consequently, ion motion arises across the amorphous area, which is aided by the segmental motion of the polymer chains [17]. The use of potassium complexed electrolyte films has been discovered to have some benefits over their lithium counterparts. Nadimicherla et al. [18] reported that the smaller ions such as ($Li^+$ and $Mg^{2+}$) possess lower mobility compared to the larger cations of ($K^+$ and $Zn^{2+}$) in polymer-based electrolytes. The smaller cations are entrenched or captured by the polymeric network. Furthermore, lithium–ion interactions with the polar polymer chains are stronger than potassium ions, and thus lithium–ion transport involves higher activation energy of 97.4 kJ $mol^{-1}$ [19]. The aim of this study is to prepare an SPE film using a biopolymer of MC doped with various concentrations of potassium iodide (KI) as the ionic source for application in EDLC device. We have investigated the effect of different KI concentration has on the conductivity of MC. Also, the electrolyte with the highest conductivity was employed in the EDLC and its decomposition potential and specific capacitance were investigated. Figure 1 depicts the schematic diagram of an EDLC cell. As seen in Figure 1, the electrolyte is inserted between two activated carbon (AC) electrodes and then packed in coin cells of CR2032 to fabricate the EDLC. The prepared EDLC device was sandwiched in a Teflon case holder with two stainless steel electrodes to investigate the capacitive behavior of the device. While measuring the impedance data of the films, the arrangement of the cell was stainless steel electrolyte film stainless steel.

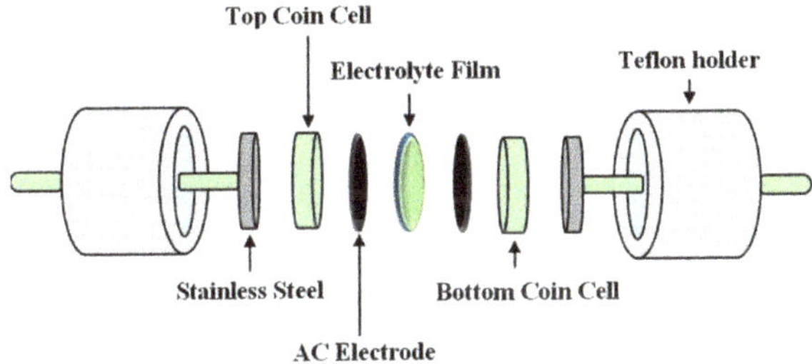


**Figure 1.** The schematic diagram of an EDLC cell. Adapted from reference [4].

## 2. Experimental Details

### *2.1. Materials and Electrolyte Preparation*

MC powder was used as a host polymeric raw material and KI salt was used as the ionic source. Both reagents were purchased from Sigma-Aldrich (Kuala Lumpur, Malaysia). The electrolytes were prepared using a solution casting technique by dissolving 1 g MC in 50 mL distilling water, with constant stirring, at room temperature for ~3 h. Subsequently, various amounts of KI salt were added to the MC solutions separately. The solutions were stirred continuously until a homogenous polymer–salt complex was obtained. The quantity of salt was varied from 10 to 40 weight percent (wt %) in steps of ten to obtain MC-KI electrolytes. The electrolyte samples were correspondingly specified as MCKI0, MCKI1, MCKI2, MCKI3, and MCK4 for MC incorporated with 0, 10, 20, 30, and 40 wt % of KI. The choice of KI concentrations is based on the ability of the MC to accommodate and dissolve the salt. Eventually, the solutions were cast on four individual categorized glass Petri dishes and left at room temperature to slowly evaporate the solvent. The films were further dried by transferring the prepared films to a desiccator.

### 2.2. Impedance Spectroscopy and FTIR Study

Electrical impedance spectroscopy (EIS) at the SPE was conducted using a Z HI-tester (Nagano, Japan) at a DC potential was 0.04 V, onto which an Ac voltage of peak-to-peak amplitude 10 mV was superimposed, over a frequency range of 5 MH and 50 HZ.

The inductance-capacitance-resistance (LCR) meter (Z HI-tester) was used to study the solid polymer electrolyte's electrical impedance spectroscopy (EIS) in the frequency range of (50 Hz $\leq f \leq$ 5 MHz). The DC potential was 0.04 V. An SPE film of geometric area of 2.01 $cm^2$ was kept between two stainless-steel electrodes by applying a spring pressure which is used to press the electrolyte films. The stainless-steel electrode was used as the working, reference, and counter electrodes while the reference and counter electrodes were combined together. The EIS data were fitted with the electric equivalent circuit (EEC) model. The common electrical elements such as resistors and capacitors are used in this model. The EEC model is simple method and provides the entire picture of the system [5].

A spotlight 400 Perkin-Elmer spectrometer (Malvern Panalytical Ltd., Malvern, UK) was employed to perform the Fourier Transforms Infrared (FTIR) spectroscopy measurements. The transmitting range was performed between 940 and 4000 $cm^{-1}$ with a resolution of 2 $cm^{-1}$.

It is vital to use Equation (1) to measure the DC ionic conductivity ($\sigma_{dc}$) of the MCKI samples based on the bulk resistance ($R_b$) value [20,21]

$$\sigma_{dc} = \left(\frac{1}{R_b}\right) \times \left(\frac{t}{A}\right) \tag{1}$$

where $t$ and $A$ denote the sample thickness and electrode area, respectively. The dielectric constant ($\varepsilon'$) and dielectric loss ($\varepsilon''$) are obtained using Equations (2) and (3) [20,21].

$$\varepsilon' = \frac{Z''}{(Z'^2 + Z''^2)C_o\omega} \tag{2}$$

$$\varepsilon'' = \frac{Z\prime}{(Z'^2 + Z''^2)C_o\omega} \tag{3}$$

where, $\omega$ and $C_o$ denote the angular frequency and capacitance, which are given by ($\omega$ = $2\pi f$) and $\varepsilon_o A/t$, respectively, where $\varepsilon_o$ stands for the free space permittivity, $A$ the electrode area and $t$ the thickness of the film [22].

The real and imaginary ($M_i$ and $M_r$) parts of complex electric modulus ($M$*) were calculated using Equations (4) and (5) [23,24].

$$M' = [\frac{\varepsilon'}{(\varepsilon'^2 + \varepsilon''^2)}] = Z''C_o\omega \tag{4}$$

$$M'' = [\frac{\varepsilon''}{(\varepsilon'^2 + \varepsilon''^2)}] = Z'C_o\omega \tag{5}$$

### 2.3. Study of Transference Number Measurement (TNM) and Linear Sweep Voltammetry (LSV)

In TNM, two types of ionic transport, $t_{ion}$ and electron transport $t_{el}$ for the most conducting sample (MCKI4) were studied. A DP3003 digital DC power supply (V & A instrument, Shanghai, China) was employed to polarize the cell against time at room temperature by applying a working voltage of 0.2 V. Linear sweep voltammetry (LSV) was used to determine the maximum potential window for the (MCKI4) film using a Digi-IVY DY2300 potentiostat (Neware, Shenzhen, China). The scan rate was fixed at 10 mV/s, and then the sample was sandwiched between two stainless steel electrodes with Teflon holders. Equations (6) and (7) were used to measure the transport ions ($t_{ion}$) and transport electrons ($t_{el}$) of the MCKI4 film, as the film was positioned between two stainless-steel electrodes [25].

$$t_{ion} = \frac{I_i - I_{ss}}{I_i} \quad (6)$$

$$t_{el} = 1 - t_{ion} \quad (7)$$

where $I_i$ refers to the initial current, containing ions and electrons and $I_{ss}$ stands for the current of the steady-state that contains only electrons.

### 2.4. EDLC Fabrication

Typically, the ingredients used to prepare electrodes include solvent and carbonaceous materials. In preparing the EDLC electrodes, 0.25 g of carbon black, 3.25 g of activated carbon, and 0.5 g of polyvinylidene fluoride (PVdF) were dry mixed in a planetary ball miller (XQM-0.4, Fujian, China) at 500 rpm for ~20 min. Then, all powders were dissolved and stirred continuously in 20 mL of N-methyl pyrrolidone until it became a dark black solution. In the next step, the black solution was covered by an aluminum foil using a doctor blade technique. Subsequently, an oven was used to dry the coated aluminum foil for a specific time at ~60 °C. To eliminate any excess moisture, the electrodes were placed in a silica gel desiccator. The relatively uppermost conducting sample was located between a pair of activated carbon electrodes and packaged in coin cells of CR2032. Eventually, in order to perform cyclic voltammetry (CV) of the assembled EDLC, the Digi-IVY DY2300 potentiostat has been employed at various scan rates of 10, 20, 50, and 100 mV/s and charged from 0 to 0.9 V. The specific capacitance, $C_s$ for the assembled EDLC has been determined using Equation (8) [25].

$$C_{cv} = \int_{V_i}^{V_f} \frac{I(V)dV}{2mv\left(V_f - V_i\right)} \quad (8)$$

where $V_i$ is the initial potential (i.e., 0 V), and $V_f$ is the final potential (i.e., 0.9 V), $m$ and $v$ are the mass of active material and the potential sweep rates (mV/s), respectively. $I(V)dV$ denotes the area under a cyclic voltammetric trace.

## 3. Result and Discussion

### 3.1. Impedance Study

Polymer electrolytes were commonly applied to devices as a part of an advanced material class. Impedance spectroscopy plays a crucial role in studying the electrical properties of a wide range of polymeric electrolyte materials. It is also a powerful technique for analyzing the ionic conductivity of new materials used in electrochemical energy systems, including EDLCs, charge transfer resistance, and diffusion layer. Plots of impedance spectra ($Z_i$ versus $Z_r$) for the MCKI1, MCKI2, MCKI3, and MCKI4 systems are shown in Figure 2a–d. In general, the impedance responses are usually characterized by a semicircle in the high frequency region and a straight line in the low frequency region [26].

EIS data are commonly analyzed by fitting to an equivalent electrical circuit model (EEC). Most of the circuit elements in the model are common electrical elements such as resistors, capacitors, and inductors. To be useful, the elements in the model should have a basis in the physical electrochemistry of the system. The EEC method has been used to investigate the EIS because it is simple and shows the entire picture of the system [5,27]. The impedance diagrams in Figure 1 can generally be represented by an equivalent circuit consisting of a charge transfer resistance ($R_b$) in a parallel arrangement with constant phase element 1 (CPE1) in high frequency region and in a series arrangement with constant phase element 2 (CPE2) in the low frequency region, as shown in the inset of Figure 1. The impedance arising from CPE, ZCPE, is expressed by Equation (9) [5,27]

$$Z_{CPE} = \frac{1}{C\omega^p}\left[\cos\left(\frac{\pi p}{2}\right) - i\sin\left(\frac{\pi p}{2}\right)\right] \quad (9)$$

Here, C is the CPE capacitance, $\omega$ is the angular frequency and $p$ is related to the EIS deviation from the imaginary axis. The $Z_r$ and $Z_i$ related to the EEC (insets of Figure 2a–d) are formulated by Equations (10) and (11)

$$Z_r = \frac{R_b^2 C_1 \omega^{p_1} cos(\pi p_1/2) + R_b}{2R_b C_1 \omega^{p_1} cos(\pi p_1/2) + R_b^2 C_1^2 \omega^{2p_1} + 1} + \frac{cos(\pi p_2/2)}{C_2 \omega^{p_2}} \tag{10}$$

$$Z_i = \frac{R_b^2 C_1 \omega^{p_1} sin(\pi p_1/2)}{2R_b C_1 \omega^{p_1} cos(\pi p_1/2) + R_p^2 C_1^2 \omega^{2p_1} + 1} + \frac{sin(\pi p_2/2)}{C_2 \omega^{p_2}} \tag{11}$$

Here, $C_1$ is the capacitance of CPE1 at the bulk of the electrolyte; $C_2$ is the CPE2 capacitance at the electrode-electrolyte interface; $p_2$ is the offset from the real axis and $p_1$ is the offset of the semicircle from the imaginary axis. The fitting parameters in the EEC are listed in Table 1. As seen in Table 1, $C_1$ and $C_2$ increased with increasing salt concentration as the number density of ions increases and they transport from the bulk of the electrolyte to the surface of the electrodes. In addition, the conductivity is also increased with increasing salt amount due to the dissociation of more salts to ions and the decrease in the $R_b$ value as seen in Figure 2a–d.

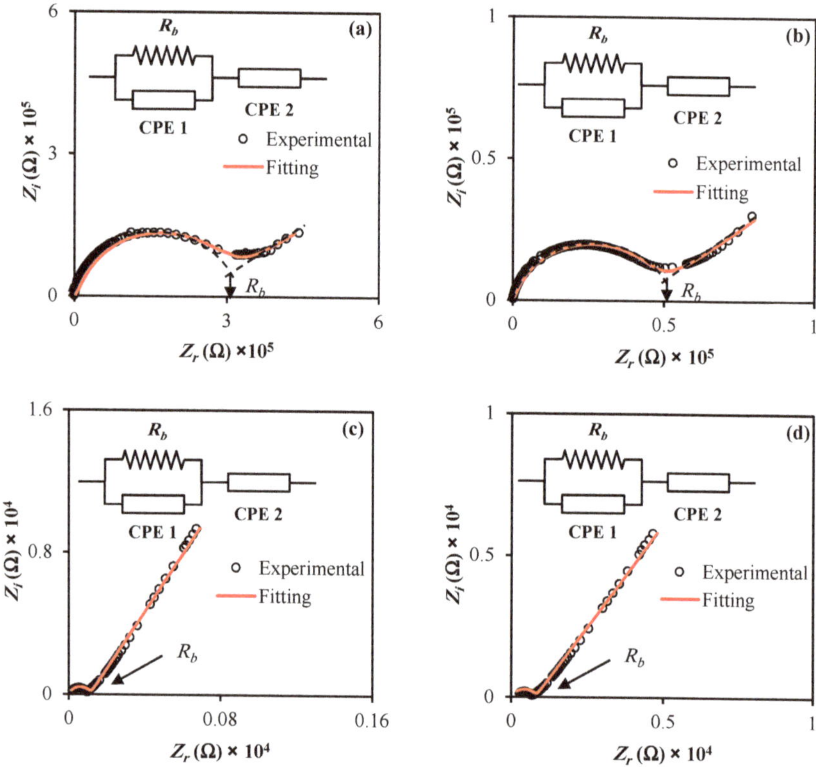


**Figure 2.** Impedance plots for (**a**) MCKI1, (**b**) MCKI2, (**c**) MCKI3, and (**d**) MCKI 4 electrolyte films.

**Table 1.** The EEC fitting parameters for the systems fabricated.

| Sample | $P_1$ (rad) | $P_2$ (rad) | $C_1$ (F) | $C_2$ (F) |
|---|---|---|---|---|
| MCKI1 | 0.90 | 0.41 | $2 \times 10^{-10}$ | $3.33 \times 10^{-7}$ |
| MCKI2 | 0.87 | 0.42 | $4 \times 10^{-10}$ | $1.43 \times 10^{-6}$ |
| MCKI3 | 0.76 | 0.65 | $6.67 \times 10^{-9}$ | $2.44 \times 10^{-6}$ |
| MCKI4 | 0.73 | 0.62 | $1.11 \times 10^{-8}$ | $4.55 \times 10^{-6}$ |

The electrode polarization is responsible for appearing the spike in Figure 2a–d at the interfaces between electrodes and electrolytes owing to blockage of ions at the electrode-electrolyte interfaces. Consequently, the electrode polarization outcome is caused by the formation of an electric double layer, resulting in free charge accumulation at the interfaces between electrodes and electrolytes. The linear increase in impedance in low frequency region in Figure 2 is expected to be a straight line (90 degree) parallel to the imaginary axis. However, there is an inclination by nearly 45° from the straight line due to the electrode polarization which causes to block of ions at the surface of the electrodes as seen in Figure 2a–d. Notably, the semicircular feature in the high frequency region has significantly diminished as KI was increased to 30 wt % and 40 wt %.

Equation (1) is used to compute the dc ionic conductivity by measuring the sample thickness and $R_b$ and the conductivity values are summarized in Table 2. As seen in Table 2, the dc conductivity increased when concentration of salt increased as more ions formed at higher salt concentration. From Equation (1), the lowest $R_b$ value shows the highest ionic conductivity [28]. It can be noted that the bulk resistance decreases with increasing the KI salt concentrations from 10 to 40 wt %. $\mu$ is related to the number density ($n$) and electrolyte conductivity ($\sigma_{dc}$) by Equation (12) [29]

$$\sigma_{dc} = ne\mu \tag{12}$$

where, $n$ is the density of the charge carrier, $\mu$ denotes mobility of ions, and $e$ denotes an electronic charge. It was established that the polymer electrolytes must have a dc ionic conductivity in the range between $10^{-3}$ and $10^{-5}$ S cm$^{-1}$ in order for it to be used in electrochemical devices [25,30,31]. Researchers have discovered that the conductivity value in this range is desirable for use in energy devices [25,30,31]. Shuhaimi et al. [32] were obtained the highest conductivity of $2.1 \times 10^{-6}$ S cm$^{-1}$ for the system of MC-$NH_4NO_3$ based biopolymer electrolyte.

**Table 2.** Numerical values of $\sigma_{dc}$, $D$, $\mu$, and $n$ at ambient temperature.

| Sample | $\sigma_{dc}$ (S cm$^{-1}$) | $D$ (cm$^2$ s$^{-1}$) | $\mu$ (cm$^2$ V$^{-1}$ s) | $n$ (cm$^{-3}$) |
|---|---|---|---|---|
| MCKI1 | $4.65 \times 10^{-8}$ | $1.15 \times 10^{-9}$ | $4.48 \times 10^{-8}$ | $6.49 \times 10^{18}$ |
| MCKI2 | $3.59 \times 10^{-7}$ | $1.35 \times 10^{-9}$ | $5.26 \times 10^{-8}$ | $4.25 \times 10^{19}$ |
| MCKI3 | $1.35 \times 10^{-5}$ | $2.00 \times 10^{-9}$ | $7.78 \times 10^{-8}$ | $1.08 \times 10^{21}$ |
| MCKI4 | $1.93 \times 10^{-5}$ | $2.13 \times 10^{-9}$ | $8.29 \times 10^{-8}$ | $1.45 \times 10^{21}$ |

As all the impedance data composed of a semicircular feature and a linear impedance, transport parameters including $D$, $\mu$ and $n$ of ions are determined using the following equations [26,28]. The $D$ of the ions is calculated using Equation (13),

$$D = \frac{(K_2\varepsilon_o\varepsilon_r A)^2}{\tau_2} \tag{13}$$

where $\varepsilon_r$ is the dielectric constant, $\tau_2$ is the reciprocal of angular frequency, which corresponds to the lowest value of $Z_i$.

The $\mu$ of the ions is determined using Equation (14)

$$\mu = \left[\frac{eD}{K_B T}\right] \tag{14}$$

where $T$ is the absolute temperature and $K_b$ is the Boltzmann constant.

Since the $\sigma_{dc}$ is given by Equation (12), the number density of ions ($n$) is calculated using Equation (15)

$$n = \left[\frac{\sigma_{dc} K_b T \tau_2}{(eK_2 \varepsilon_o \varepsilon_r A)^2}\right] \tag{15}$$

Table 2 lists the ion transport parameters for each electrolyte system.

Based on Table 2, the $D$ increased as the KI concentration increased from 10 to 40 wt %. The identical tendency is seen by $\mu$ as listed in Table 2 where $\mu$ increased. The increase of $\mu$ and $D$ is related to the increase of chain flexibility with the existence of slat [28]. Consequently, an improvement of conductivity is resulted.

Figure 3a,b show the Bode plot for each electrolyte film at room temperature. An earlier study [33] indicated that the capacitive region is a plateau region between $10^{-2}$ Hz and 100 Hz. However, this feature is not observed in Figure 3 because of the limitation of frequency of our measuring equipment. As described at the EIS plots, the semicircle is associated with ion transfer in the electrolyte and the linear feature arises from ions diffusion and therefore their accumulation at the interfaces between electrode and electrolyte [33] which leads to an electrical double-layer capacitances. It was shown that, by increasing the amount of salt from 10 wt % to 40 wt %, the linear feature increased and the resistance reduced from $3.3 \times 10^5$ to $8 \times 10^2$ Ω, because of the more carrier density. As seen in Figure 3a the electrolyte film has high charge transfer resistance ($R_{ct}$) while with increasing salt the $R_{ct}$ decreased as shown in Figure 3b. The dispersion region between 40 Hz and 40,000 Hz is ascribed to the phenomena of ion diffusion and the high-frequency region is ascribed to the $R_{ct}$. In Figures 2 and 3, it is seen that the sample loaded with 40 wt % of KI has the lowest $R_{ct}$ and hence a large conductivity resulted. Therefore, the Bode plot supports the result measured from the impedance study.

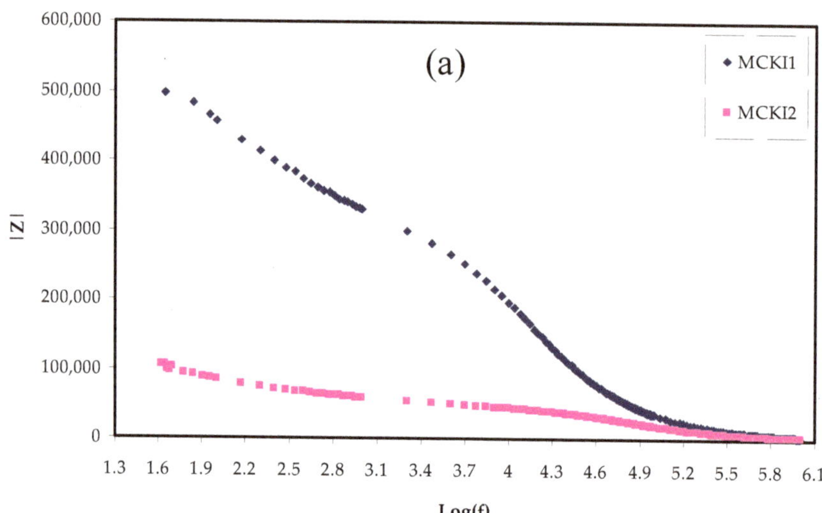


**Figure 3.** *Cont.*

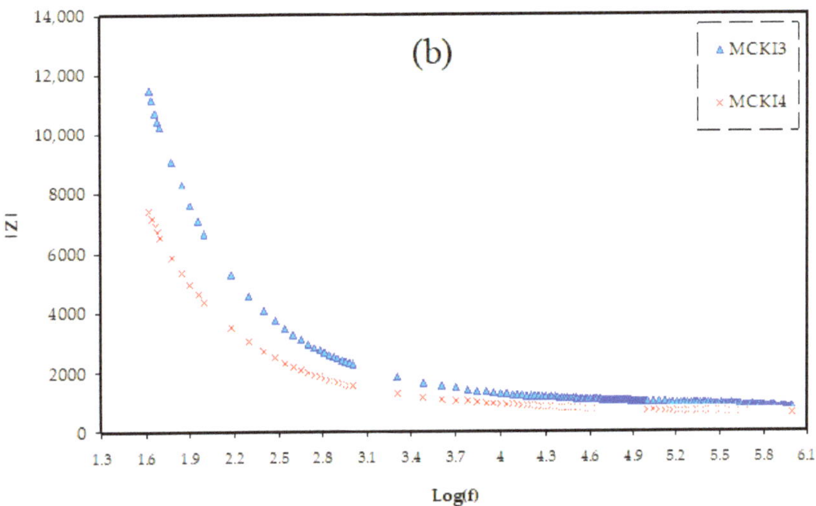


**Figure 3.** Bode plots for (**a**) MCKI1 and MCKI2 and (**b**) MCKI3 and MCKI4 electrolyte samples.

*3.2. Dielectric Properties*

Complex electric modulus, defined as the inverse of complex relative permittivity, can be a significantly powerful tool for analyzing dielectric behavior of a polymeric insulating material, especially at relatively high temperatures, where complex permittivity usually becomes very high due to electrode polarization and carrier transport. The core of electrochemical devices are ions conducting solid electrolytes, and its electrical properties investigation such as $\sigma_{dc}$, $\varepsilon^*$, and electric modulus ($M^*$) are essential to understanding the ions transport process [21]. The real part ($\varepsilon'$) is related to ion storage efficiency or polarizing ability, while the imaginary part ($\varepsilon''$) is the necessary energy for dipole alignment [34]. The $\varepsilon'$ and $\varepsilon''$ are determined using Equations (2) and (3).

Figure 4a,b display the frequency dependency of the $\varepsilon'$ and $\varepsilon''$ for the MC polymer incorporated with various concentrations of KI salt. It can be noted that the system integrated with 40 wt % of KI has the highest dielectric constant at a low-frequency region. It might be owing to the electrode polarization and also space charge effects. The rise in dielectric constant can be explained by the high charge carrier concentration of the system and its amorphous composition [35]. It is seen that as the salt content (KI) increases, the $\varepsilon'$ and $\varepsilon''$ increase. This is in agreement with the increase in number density and mobility of ions when the KI content increased as shown in Table 2. Both of the $\varepsilon'$ and $\varepsilon''$ values are elevated at low frequencies and decreased as frequency rises, indicating polarization effect due to charge accumulations near electrodes at low frequency and dipoles do not obey the field variation at a high dispersion frequency region [36]. The dielectric values remain stable at high-frequency regions due to the interfaces of the electrode–electrolyte become marginal as the frequency increases. The decreased value of both $\varepsilon'$ and $\varepsilon''$ with increasing frequency means that the electrolyte films are non-Debye behavior [37].

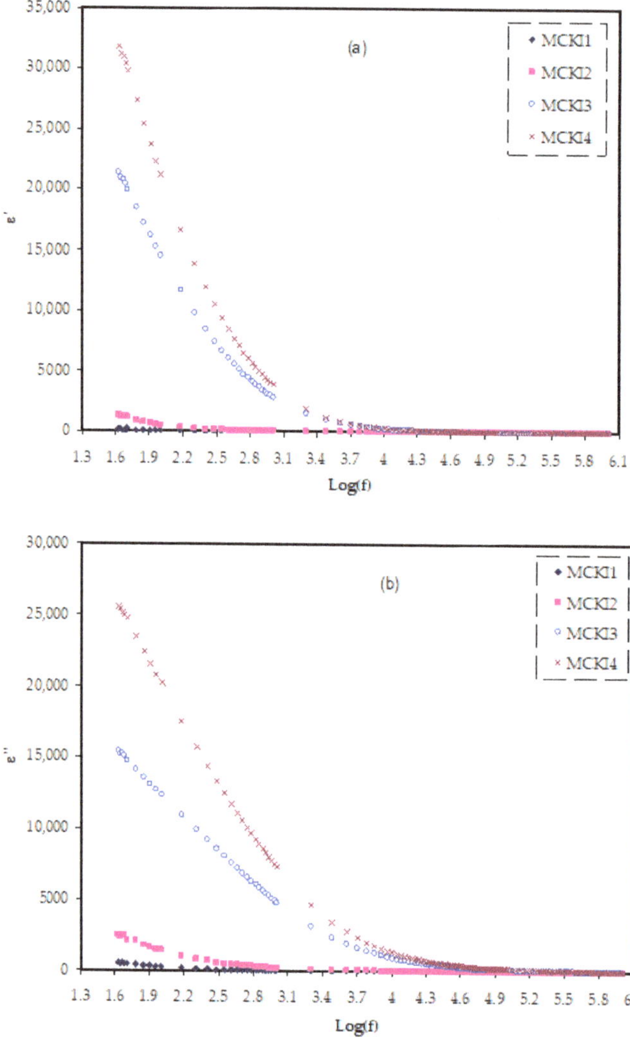


**Figure 4.** Dielectric plot for (**a**) $\varepsilon'$ and (**b**) $\varepsilon''$ variation against frequency for the MCKI samples.

The $Z_r$ and $Z_i$ data were achieved from the EIS data and then used to determine the $\varepsilon'$ and $\varepsilon''$ data. The $\varepsilon'$ and $\varepsilon''$ were used to find the $tan\ \delta$. The $tan\ \delta$ is the ratio between energy disperse and energy stored in a periodical field which is also called dissipation factor [23] and it is determined using Equation (16).

$$tan\ \delta = \frac{\varepsilon''}{\varepsilon'} \quad (16)$$

Dielectric loss is the energy dissipation by the transfer of charges in an alternating electric field as polarization switches direction. When the electric field is applied, polarization happens and charges are moved relative to the electric field. Dielectric loss causes a decrease in the overall electric field. The total amount of polarization that can happen in a dielectric relies on the molecular symmetry of the insulator material and is known as dipole moment. The influence of the dipole moment in a dielectric material is called loss tangent.

The ratio of $\varepsilon''$ to $\varepsilon'$ is defined as *tan* $\delta$, where $\delta$ denotes a loss angle. The *tan* $\delta$ is determined using the relation below [23]. Loss tangent (*tan* $\delta$) was further investigated for the MC polymer incorporated with various KI concentrations. Figure 5 shows the loss tangent (*tan* $\delta$) spectra versus frequency at room temperature. The relation between loss tangent and frequency reveals some interesting behavior. Overall, the loss tangent increases with increasing the applied frequency due to the domination of the Ohmic components. It reaches a high value at a certain frequency, and followed by decreases at a high frequency, owing to the increasing nature of the reactive components [38]. Notably, MCKI4 displays the highest shift to the high frequency and the maximum value relative to the other samples due to the value of dielectric constant $\varepsilon'$ for the MCKI4 as shown in Figure 4a [39]. The presence of the peaks at a characteristic frequency can be argued for indicating the presence of dipole relaxation in the electrolytes. It has been reported that improving the segmental motion of polymer chains decreases the relaxation time, allowing the transport process easier. This is expressed mathematically as $\tau = 1/2\pi f_{max}$, where $\tau$ is the ionic charge carrier's relaxation time [40].

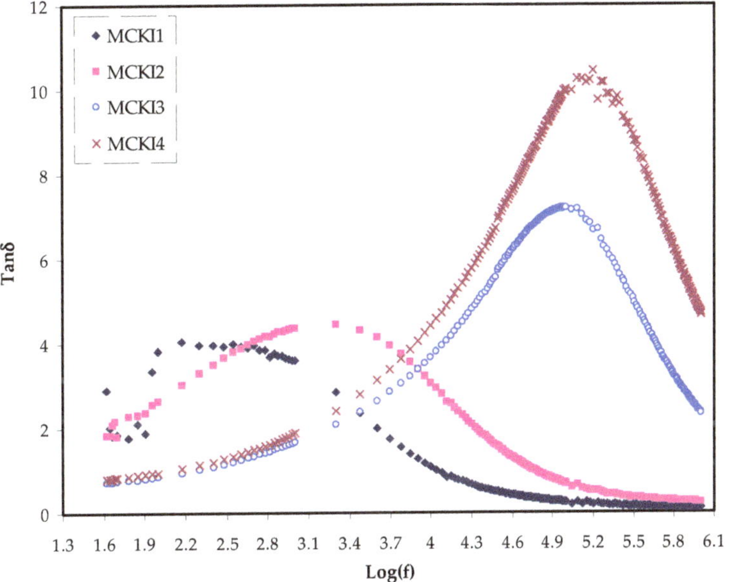


**Figure 5.** (*tan* $\delta$) spectra versus frequency at room temperature for the MCKI electrolytes.

The real, $M_r$ and imaginary, $M_i$ components of the electric modulus $M^*$ against frequency for the MCKI based solid polymer electrolytes are shown in Figures 6 and 7, respectively. The $M'$ and $M''$ are determined using Equations (4) and (5).

From the figures, $M_r$ values are noted to decrease with decreasing frequencies until they reach zero, meaning that the polarization was eliminated. Therefore, the $M_r$ values rise with increasing frequency and at the highest frequency, the maximum $M_r$ was obtained. This could be attributed to the fact that the relaxation process occurs at various frequency values [41]. The observed dispersion is essentially as of conductivity relaxation covering several frequencies, indicating the presence of $\tau$ that has to occur with a loss peak in the figure of the imaginary part of the dielectric modulus versus frequency. As $M_i$ has clearly a lower value at a low frequency, this may be attributed to the higher capacitance coupled with the polarization effect. No peak is present in Figure 6 along with its entire frequency range. It could be referring to the $M_r$ which is equivalent to the $\varepsilon'$ in the $\varepsilon^*$ representation, which $M_r$ shows the material's potential for energy conversion [42].

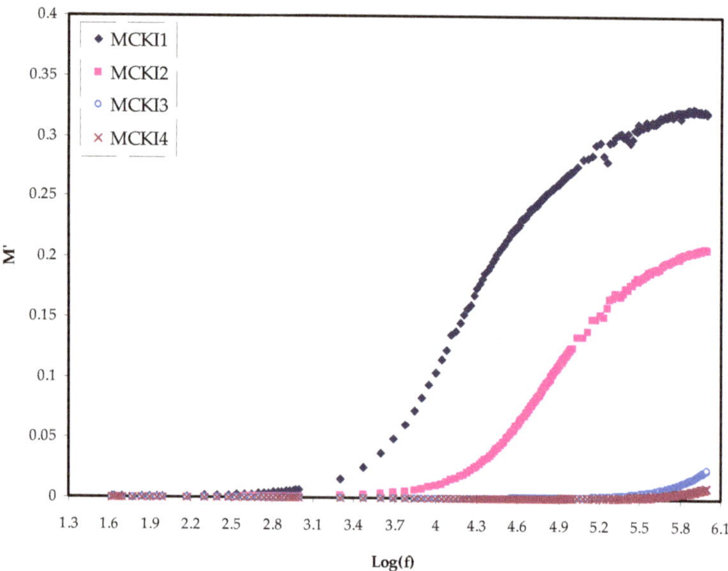


**Figure 6.** Electric modulus plot of $M_r$ against log(f)for the MCKI samples.

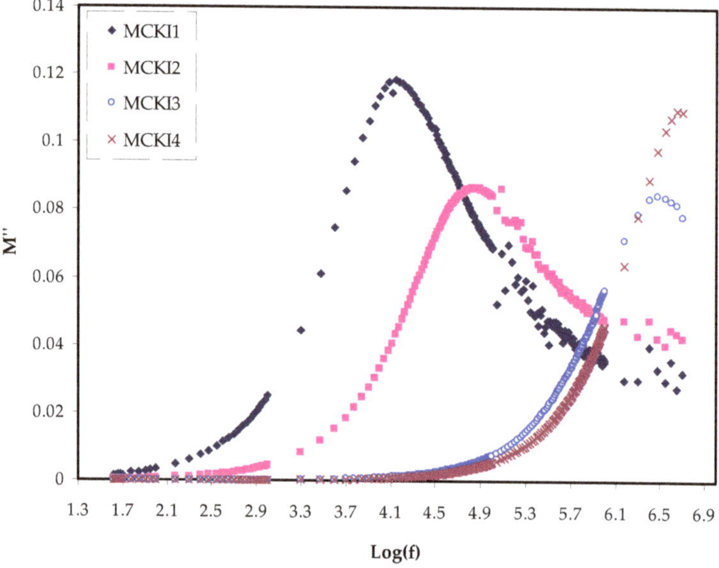


**Figure 7.** Electric modulus plot of $M_i$ against log(f) for the MCKI samples.

### 3.3. FTIR Study

The technique of FTIR spectroscopy has been used to investigate the interactions between ions and atoms of the MCKI electrolytes. Also, such interactions can lead to the changes in the vibration modes of the polymer electrolyte. The FTIR spectra of the pure MC and MCKI based solid polymer electrolyte over the wavenumber range of 940–4000 cm$^{-1}$ are displayed in Figure 8a,b. The broad peak observed at around 1050 cm$^{-1}$ corresponds to the antisymmetric stretch of an asymmetric oxygen bridge in its cyclohexane ring of pure

MC. The water contamination from the KI salt causes a broad peak at 3400 $cm^{-1}$ of the O-H stretching band. The observed peak intensity changes as the weight percent of KI salt was increased from 0 to 40 % in the MC-KI electrolyte systems, as shown in Figure 8a,b [43,44]. A peak that appears in the wavenumber region of 2800–2950 $cm^{-1}$ is corresponding to the C-H stretching mode of methylcellulose. Through the inclusion of KI salt, the peak seems to shift slightly from 2850 $cm^{-1}$ to 2990 $cm^{-1}$. This shift of the peak may be an indication of the complexation of $K^+$ cation and the MC host polymer. However, the slight change in the C-O ether bands indicates that the complexation did not considerably modify the molecular structure of the MC host polymer. Furthermore, the change in peak intensity with increasing KI concentrations supports that the presence of KI salt in the system has a significant impact on the conductivity of the MCKI electrolyte systems [45].

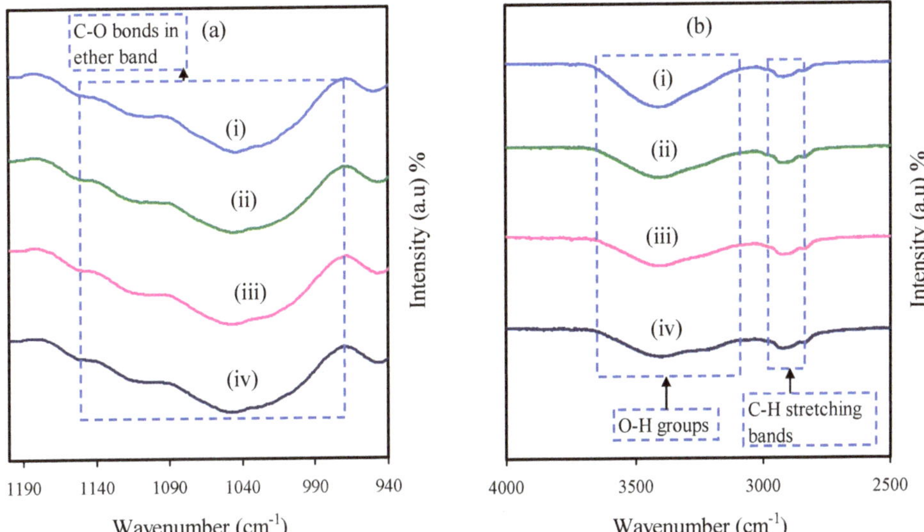


**Figure 8.** FTIR spectra of the MCKI samples at a wavenumber of (**a**) 940–1200 $cm^{-1}$ and (**b**) 2500–4000 $cm^{-1}$ for (i) MCKI1 (ii) MCKI2, (iii) MCKI3, and (iv) MCKI4 electrolyte samples.

### 3.4. EDLC Study

#### 3.4.1. Study of the TNM

Both ions and electrons in polymer electrolytes are generally responsible for their conductivity. Through this technique, the dominant charge carrier in the polymer electrolyte can be evaluated [46]. Figure 9 shows the current versus time plot, obtained by dc polarization at 0.2 V, for the $MCK1_4$ film. Equations (6) and (7) were used to determine the $t_{ion}$ and $t_{el}$ of the MCKI4 film.

According to Figure 9, the initial total current was found to be 22 μA [47]. Therefore, a large drop is observed over time until being constant in a completely depleted case due to the transport of ionic species from the bulk of the MCKI4 electrolyte to the electrode-electrolyte interfaces. When the cell reaches the steady state, it is polarized, and the residual current is only carried by electrons due to the stainless-steel electrodes block both cations and anions while allowing only electrons to move through it. In this analysis, the measured $t_{el}$ value was 0.12 and the $t_{ion}$ was found to be 0.88, which is close to an ideal value of 1 [28], indicating that ions in the MCKI4 film is the majority charge carrier [48]. The finding obtained in this work is comparable with the $t_{ion}$ value of 0.86 as reported by Aziz et al. for the polymer electrolyte system of chitosan: dextran: $NH_4Br$ [49].

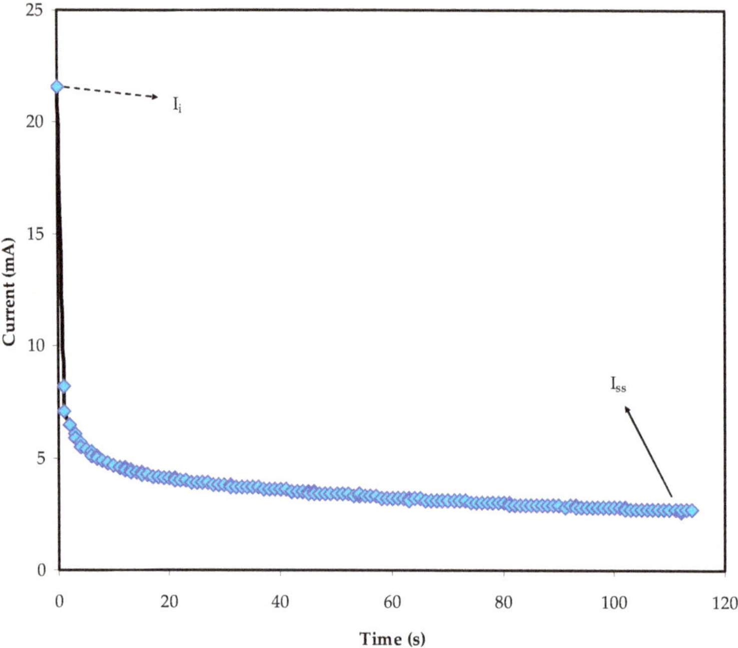


**Figure 9.** DC polarization curve of current versus time for the MCKI4 sample.

### 3.4.2. LSV Study

The potential stability of the polymer electrolyte systems needs to be established for energy device research. The absolute potential limit of the electrolytes can be computed in terms of linear sweep voltammetry LSV examination [50]. The LSV for the most conducting sample MCKI4 at 10 mV/s is shown in Figure 10, in which the potential was scanned from 0 to 2.5 V. When potential approaches to 1.8 V, the electrolyte reaches decomposition voltage as revealed by a significant increase in current values. Also, there is no evidence of a redox reaction occurring within the potential window until 1.8 V. Based on the previous study, the electrolyte with the potential window of 1.8 V is sufficient to be used for application in proton energy devices [51]. Other research findings relating to MC-based biopolymer electrolytes are comparable to this work. According to Kadir et al. [52], MC-based electrolytes displayed a decomposition voltage of 1.53 V when $NH_4Br$ and glycerol were used as the ionic source and plasticizer, respectively. The breakdown potential of 1.9 V was reported for the biopolymeric system of starch-chitosan-$NH_4I$ with the existence of glycerol [53], which is similar to this study.

### 3.4.3. Cyclic Voltammetry (CV) Study

CV as an insightful technique can be employed to examine the EDLCs in terms of both qualitative and quantitative features [54]. It is used to further evaluate the efficiency of the MCKI4 electrolyte in the construction of the EDLC. The CV responses of the MCKI4 electrolyte at various scan rates of 10, 20, 50, and 100 mV/s are shown in Figure 11 in the potential range of 0 to 0.9 V.

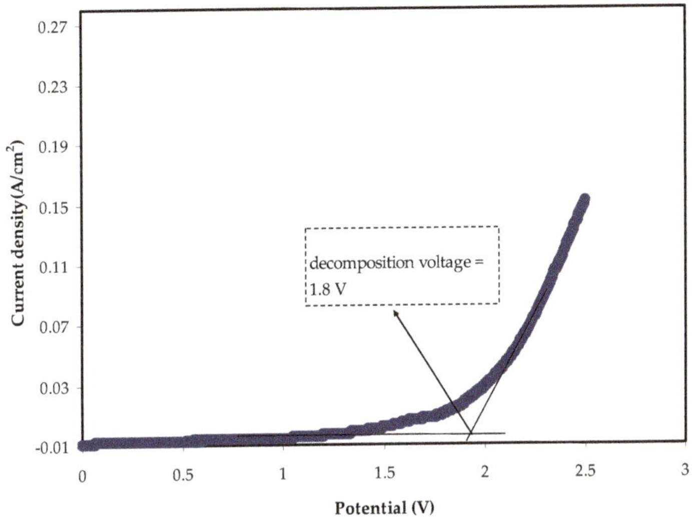


**Figure 10.** Current versus potential for the highest conducting (MCKI4) sample.

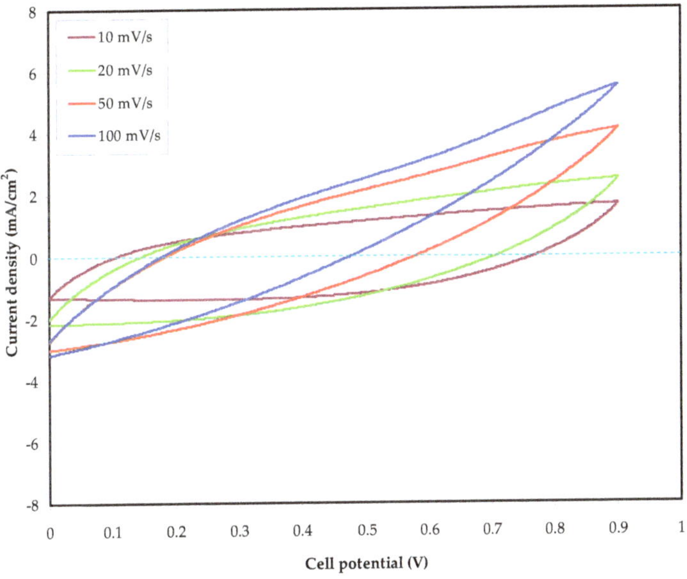


**Figure 11.** Cyclic voltammograms for the assemble EDLC in the potential range of 0 to 0.9 V.

The CV response has a rectangular form, indicating that the current is independent of the potential. However, the shape of the cyclic voltammogram (CV) deviates from the rectangular shape when the scan rate increases [55]. The CV in Figure 11 showed that the EDLC exhibits a capacitive behavior, indicating that the system of the energy storage is a non-Faradaic mechanism. In this process, the charge stored in the EDLC system comes from ion accumulation at the electrode/electrolyte interfaces. As a consequence, ion accumulation and adsorption occur in the place of deintercalation and intercalation via a non-Faradaic mechanism. In addition, ions from the bulk of the electrolyte form a charge double-layer, which then saves potential energy [56,57]. Notably, the CV displays a leaf-like shape with no redox peaks. The CV profile revealed a little divergence from its rectangular form at higher scan rates, which may be due to the porosity of the electrodes as well as internal resistance. The porosity of the carbon electrodes induces a relatively high internal resistance, which causes the CV to appear leaf-like in shape [58]. Since the CV possesses no redox peaks, it is reasonable to infer that a quick Faradaic reversible reaction has not occurred [59].

The specific capacitance ($C_s$)are determined using Equation (8) by measuring the area of the CV profile, mass of the activated carbon electrode, scan rate, and the initial and final values of applied voltage. The measured specific capacitance values, $C_s$ using CV curves for the assembled EDLC at different scan rates are shown in Table 3 and Figure 12. The calculated $C_s$ value of 113.39 F/g at the sweep rate of 10 mV/s decreased to 11.84 F/g at 100 mV/s. The low $C_s$ value at high scan rates is attributed to the high energy loss caused by the decrease in the density of stored charges, which results in a lower $C_s$ value [60]. Table 4 displays the measured $C_s$ value of the EDLC for several systems based on solid biopolymer electrolytes mentioned in the literature. Interestingly, the $C_s$ value obtained in this work is high and comparable to some of these results.

**Table 3.** Specific capacitance ($C_s$) of the EDLCs using CV curves.

| Scan Rates (mV/s) | Specific Capacitance, $C_s$ F/g |
|---|---|
| 10 | 113.39 |
| 20 | 69.16 |
| 50 | 27.48 |
| 100 | 11.84 |

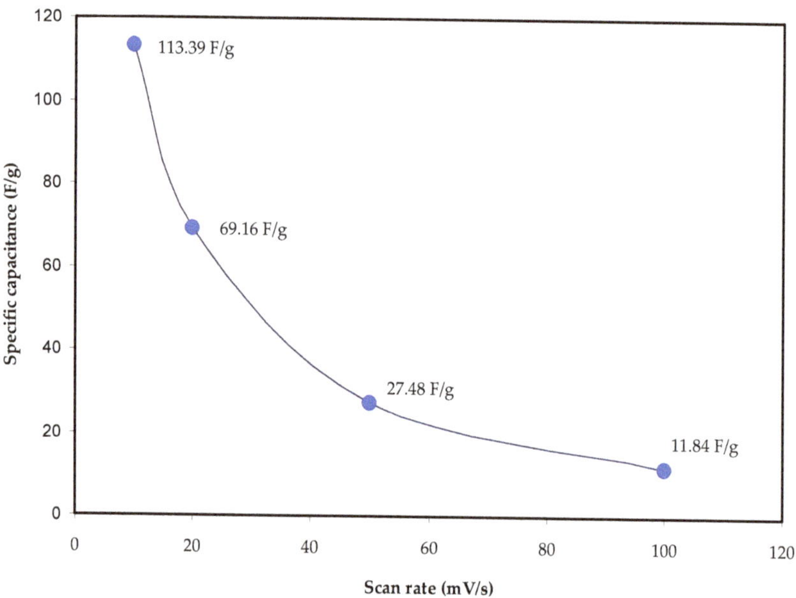


**Figure 12.** The calculated specific capacitance, $C_s$ for the assembled EDLC at a different scan rate.

**Table 4.** Specific capacitance ($C_s$) of the EDLCs using different polymer electrolytes at room temperature.

| Biopolymer Electrolytes | Specific Capacitance, $C_s$ F/g | Scan Rates (mV/s) | Reference |
|---|---|---|---|
| Chitosan-PVA-$Mg(CF_3SO_3)_2$:glycerol | 32.69 | 10 | [3] |
| Starch-$LiClO_4$ | 8.7 | 10 | [7] |
| MC-$NH_4NO_3$-PEG | 38 | 1 | [45] |
| MC-chitosan-$NH_4SCN$ | 66.3 | 10 | [33] |
| Carboxymethyl cellulose-$NH_4NO_3$ | 1.8 | Not stated | [61] |
| MC-chitosan-$NH_4I$-glycerol | 9.97 | 10 | [62] |
| Cellulose acetate-$LiClO_4$ | 90 | 10 | [63] |
| Chitosan-$NH_4Br$-glycerol | 7.5 | 10 | [64] |
| MC-Starch-$LiClO_4$-glycerol | 45.8 | 10 | [65] |
| MC-KI | 113.39 | 10 | This work |

## 4. Conclusions

In conclusion, a biopolymer-based electrolyte using methylcellulose (MC) incorporated with various content of potassium iodide (KI) salt is crucial for EDLC device applications. The EIS outcome shows that the resistance of the transfer of charge at the bulk of the electrolyte reduced from $3.3 \times 10^5$ Ω to $8 \times 10^2$ Ω with KI concentration increased from 10 wt % to 40 wt % due to an increase in the charge carrier density. The highest conductivity of $1.93 \times 10^{-5}$ S/cm was obtained for the electrolyte doped with 40 wt % of KI. The dielectric analysis further verified the conductivity trends. The results from the FTIR spectra indicated that the complexation between ($K^+$) cation and (MC) host polymer has occurred through intensity variations of bands. TNM measurements stated that the ions were the dominant charge carrier, as the ($t_{ion}$) was identified to be 0.88. LSV analysis showed that the most conducting sample has an electrochemical stability window up to 1.8 V, verifying the suitability of the electrolyte for EDLC application. The CV response displayed its capacitance behavior, where no visible redox peak has appeared. A relatively high value of the specific capacitance $C_s$ (113.39 F/g) was obtained at the scan rate of 10 mV/s.

**Author Contributions:** Conceptualization, formal analysis, writing—original draft, methodology, supervision, S.B.A., M.M.N. and J.M.H.; Formal analysis, A.S.F.M.A. and M.A.B.; Investigation, M.A.B., M.M.N., A.M.A.; Methodology, S.B.A. and M.A.B.; Project administration, S.B.A. and M.F.Z.K.; Validation, E.M.A.D., M.A.B., A.M.A. and M.F.Z.K.; Writing—original draft, M.M.N., J.M.H., S.B.A. and A.S.F.M.A.; Writing—review & editing, S.B.A., M.A.B., M.F.Z.K. and E.M.A.D. All authors have read and agreed to the published version of the manuscript.

**Funding:** The authors would like to acknowledge the support of Prince Sultan University for paying the Article Processing Charges (APC) of this publication and for their financial support.

**Institutional Review Board Statement:** Not applicable.

**Informed Consent Statement:** Not applicable.

**Data Availability Statement:** The data presented in this study are available on request from the corresponding author.

**Acknowledgments:** We would like to acknowledge all support for this work by the University of Sulaimani, Charmo University, Prince Sultan University and Komar University of Science and Technology.

**Conflicts of Interest:** The authors declare no conflict of interest.

## References

1. Sohaimy, M.I.H.A.; Isa, M.I.N.M. Natural inspired carboxymethyl cellulose (Cmc) doped with ammonium carbonate (ac) as biopolymer electrolyte. *Polymers* **2020**, *12*, 2487. [CrossRef] [PubMed]
2. Hema, M.; Selvasekarapandian, S.; Arunkumar, D.; Sakunthala, A.; Nithya, H. FTIR, XRD and ac impedance spectroscopic study on PVA based polymer electrolyte doped with NH4X (X = Cl, Br, I). *J. Non-Cryst. Solids* **2009**, *355*, 84–90. [CrossRef]

3. Aziz, S.B.; Brza, M.A.; Dannoun, E.M.A.; Hamsan, M.H.; Hadi, J.M. The Study of Electrical and Electrochemical Properties of Magnesium Ion Conducting CS: PVA Based Polymer Blend Electrolytes: Role of Lattice Energy of Magnesium Salts on EDLC Performance. *Molecules* **2020**, *25*, 4503. [CrossRef]
4. Aziz, S.B.; Asnawi, A.S.; Abdulwahid, R.T.; Ghareeb, H.O.; Alshehri, S.M.; Ahamad, T.; Hadi, J.M.; Kadir, M. Design of potassium ion conducting PVA based polymer electrolyte with improved ion transport properties for EDLC device application. *J. Mater. Res. Technol.* **2021**, *13*, 933–946. [CrossRef]
5. Aziz, S.B.; Brevik, I.; Hamsan, M.H.; Brza, M.A.; Nofal, M.M.; Abdullah, A.M.; Rostam, S.; Al-Zangana, S.; Muzakir, S.K.; Kadir, M.F.Z. Compatible solid polymer electrolyte based on methyl cellulose for energy storage application: Structural, electrical, and electrochemical properties. *Polymers* **2020**, *12*, 2257. [CrossRef]
6. Taghizadeh, M.T.; Seifi-Aghjekohal, P. Sonocatalytic degradation of 2-hydroxyethyl cellulose in the presence of some nanoparticles. *Ultrason. Sonochem.* **2015**, *26*, 265–272. [CrossRef] [PubMed]
7. Liew, C.W.; Ramesh, S.; Arof, A.K. Enhanced capacitance of EDLCs (electrical double layer capacitors) based on ionic liquid-added polymer electrolytes. *Energy* **2016**, *109*, 546–556. [CrossRef]
8. Choudhury, N.A.; Sampath, S.; Shukla, A.K. Hydrogel-polymer electrolytes for electrochemical capacitors: An overview. *Energy Environ. Sci.* **2009**, *2*, 55–67. [CrossRef]
9. Wei, L.; Sevilla, M.; Fuertes, A.B.; Mokaya, R.; Yushin, G. Polypyrrole-derived activated carbons for high-performance electrical double-layer capacitors with ionic liquid electrolyte. *Adv. Funct. Mater.* **2012**, *22*, 827–834. [CrossRef]
10. Hadi, J.M.; Aziz, S.B.; Mustafa, M.S.; Brza, M.A.; Hamsan, M.H.; Kadir, M.F.Z.; Ghareeb, H.O.; Hussein, S.A. Electrochemical impedance study of proton conducting polymer electrolytes based on PVC doped with thiocyanate and plasticized with glycerol. *Int. J. Electrochem. Sci.* **2020**, *15*, 4671–4683. [CrossRef]
11. Wang, H.; Lin, J.; Shen, Z.X. Polyaniline (PANi) based electrode materials for energy storage and conversion. *J. Sci. Adv. Mater. Devices* **2016**, *1*, 225–255. [CrossRef]
12. Zhao, X.Y.; Wu, Y.; Cao, J.P.; Zhuang, Q.Q.; Wan, X.; He, S.; Wei, X.Y. Preparation and characterization of activated carbons from oxygen-rich lignite for electric double-layer capacitor. *Int. J. Electrochem. Sci.* **2018**, *13*, 2800–2816. [CrossRef]
13. Iro, Z.S.; Subramani, C.; Dash, S.S. A brief review on electrode materials for supercapacitor. *Int. J. Electrochem. Sci.* **2016**, *11*, 10628–10643. [CrossRef]
14. Chen, T.-Y.; Vedhanarayanan, B.; Lin, S.-Y.; Shao, L.-D.; Sofer, Z.; Lin, J.-Y.; Lin, T.-W. Electrodeposited NiSe on a forest of carbon nanotubes as a free-standing electrode for hybrid supercapacitors and overall water splitting. *J. Colloid Interface Sci.* **2020**, *574*, 300–311. [CrossRef] [PubMed]
15. Lien, C.-W.; Vedhanarayanan, B.; Chen, J.-H.; Lin, J.-Y.; Tsai, H.-H.; Shao, L.-D.; Lin, T.-W. Optimization of acetonitrile/water content in hybrid deep eutectic solvent for graphene/$MoS_2$ hydrogel-based supercapacitors. *Chem. Eng. J.* **2020**, *405*, 126706. [CrossRef]
16. Hsiang, H.-I.; She, C.-H.; Chung, S.-H. Materials and electrode designs of high-performance $NiCo_2S_4$/Reduced graphene oxide for supercapacitors. *Ceram. Int.* **2021**, *47*, 25942–25950. [CrossRef]
17. Dey, A.; Karan, S.; De, S.K. Effect of nanofillers on thermal and transport properties of potassium iodide-polyethylene oxide solid polymer electrolyte. *Solid State Commun.* **2009**, *149*, 1282–1287. [CrossRef]
18. Nadimicherla, R.; Kalla, R.; Muchakayala, R.; Guo, X. Effects of potassium iodide (KI) on crystallinity, thermal stability, and electrical properties of polymer blend electrolytes (PVC/PEO:KI). *Solid State Ionics* **2015**, *278*, 260–267. [CrossRef]
19. Sagane, F.; Abe, T.; Iriyama, Y.; Ogumi, Z. $Li^+$ and $Na^+$ transfer through interfaces between inorganic solid electrolytes and polymer or liquid electrolytes. *J. Power Sources* **2005**, *146*, 749–752. [CrossRef]
20. Hamsan, H.M.; Aziz, S.B.; Kadir, M.F.Z.; Brza, M.A.; Karim, W.O. The study of EDLC device fabricated from plasticized magnesium ion conducting chitosan based polymer electrolyte. *Polym. Test.* **2020**, *90*, 106714. [CrossRef]
21. Mustafa, M.S.; Ghareeb, H.O.; Aziz, S.B.; Brza, M.A.; Al-Zangana, S.; Hadi, J.M.; Kadir, M.F.Z. Electrochemical Characteristics of Glycerolized PEO-Based Polymer Electrolytes. *Membranes* **2020**, *10*, 116. [CrossRef]
22. Nofal, M.M.; Aziz, S.B.; Hadi, J.M.; Abdulwahid, R.T.; Dannoun, E.M.A.; Marif, A.S.; Al-Zangana, S.; Zafar, Q.; Brza, M.A.; Kadir, M.F.Z. Synthesis of porous proton ion conducting solid polymer blend electrolytes based on PVA: CS polymers: Structural, morphological and electrochemical properties. *Materials* **2020**, *13*, 4890. [CrossRef]
23. Pawlicka, A.; Tavares, F.C.; Dörr, D.S.; Cholant, C.M.; Ely, F.; Santos, M.J.L.; Avellaneda, C.O. Dielectric behavior and FTIR studies of xanthan gum-based solid polymer electrolytes. *Electrochim. Acta* **2019**, *305*, 232–239. [CrossRef]
24. Hadi, J.M.; Aziz, S.B.; Nofal, M.M.; Hussen, S.A.; Hamsan, M.H.; Brza, M.A.; Abdulwahid, R.T.; Kadir, M.F.Z.; Woo, H.J. Electrical, dielectric property and electrochemical performances of plasticized silver ion-conducting chitosan-based polymer nanocomposites. *Membranes* **2020**, *10*, 151. [CrossRef] [PubMed]
25. Brza, M.A.; Aziz, S.B.; Anuar, H.; Ali, F. Structural, Ion Transport Parameter and Electrochemical Properties of Plasticized Polymer Composite Electrolyte Based on PVA: A Novel Approach to Fabricate High Performance EDLC Devices. *Polym. Test.* **2020**, *91*, 106813. [CrossRef]
26. Aziz, S.B.; Nofal, M.M.; Kadir, M.F.Z.; Dannoun, E.M.A.; Brza, M.A.; Hadi, J.M.; Abdullah, R.M. Bio-Based Plasticized PVA Based Polymer Blend Electrolytes for Energy Storage EDLC Devices: Ion Transport Parameters and Electrochemical Properties. *Materials* **2021**, *14*, 1994. [CrossRef]

27. Asnawi, A.; Aziz, S.; Brevik, I.; Brza, M.; Yusof, Y.; Alshehri, S.; Ahamad, T.; Kadir, M. The Study of Plasticized Sodium Ion Conducting Polymer Blend Electrolyte Membranes Based on Chitosan/Dextran Biopolymers: Ion Transport, Structural, Morphological and Potential Stability. *Polymers* **2021**, *13*, 383. [CrossRef]
28. Brza, M.A.; Aziz, S.B.; Anuar, H.; Alshehri, S.M.; Ali, F.; Ahamad, T.; Hadi, J.M. Characteristics of a Plasticized PVA-Based Polymer Electrolyte Membrane and $H^+$ Conductor for an Electrical Double-Layer Capacitor: Structural, Morphological, and Ion Transport Properties. *Membranes* **2021**, *11*, 296. [CrossRef] [PubMed]
29. Aziz, S.B.; Brza, M.A.; Saed, S.R.; Hamsan, M.H.; Kadir, M.F.Z. Ion association as a main shortcoming in polymer blend electrolytes based on CS: PS incorporated with various amounts of ammonium tetrafluoroborate. *J. Mater. Res. Technol.* **2020**, *9*, 5410–5421. [CrossRef]
30. Dannoun, E.M.; Aziz, S.B.; Brza, M.A.; Nofal, M.M.; Asnawi, A.S.; Yusof, Y.M.; Al-Zangana, S.; Hamsan, M.H.; Kadir, M.F.Z.; Woo, H.J. The Study of Plasticized Solid Polymer Blend Electrolytes Based on Natural Polymers and Their Application for Energy Storage EDLC Devices. *Polymers* **2020**, *12*, 2531. [CrossRef]
31. Aziz, S.B.; Marf, A.S.; Dannoun, E.M.A.; Brza, M.A.; Abdullah, R.M. The Study of the Degree of Crystallinity, Electrical Equivalent Circuit, and Dielectric Properties of Polyvinyl Alcohol (PVA)-Based Biopolymer Electrolytes. *Polymers* **2020**, *12*, 2184. [CrossRef] [PubMed]
32. Shuhaimi, N.E.A.; Teo, L.P.; Majid, S.R.; Arof, A.K. Transport studies of $NH_4NO_3$ doped methyl cellulose electrolyte. *Synth. Met.* **2010**, *160*, 1040–1044. [CrossRef]
33. Aziz, S.B.; Hamsan, M.H.; Abdullah, R.M.; Kadir, M.F.Z. A promising polymer blend electrolytes based on chitosan: Methyl cellulose for EDLC application with high specific capacitance and energy density. *Molecules* **2019**, *24*, 2503. [CrossRef] [PubMed]
34. Pritam; Arya, A.; Sharma, A.L. Selection of best composition of $Na^+$ ion conducting PEO-PEI blend solid polymer electrolyte based on structural, electrical, and dielectric spectroscopic analysis. *Ionics* **2020**, *26*, 745–766. [CrossRef]
35. Jaipalreddy, M.; Chu, P. Optical microscopy and conductivity of poly(ethylene oxide) complexed with KI salt. *Electrochim. Acta* **2002**, *47*, 1189–1196. [CrossRef]
36. Elashmawi, I.S.; Gaabour, L.H. Raman, morphology and electrical behavior of nanocomposites based on PEO/PVDF with multi-walled carbon nanotubes. *Results Phys.* **2015**, *5*, 105–110. [CrossRef]
37. Jothi, M.A.; Vanitha, D.; Nallamuthu, N.; Manikandan, A.; Bahadur, S.A. Investigations of lithium ion conducting polymer blend electrolytes using biodegradable cornstarch and PVP. *Phys. B Condens. Matter* **2020**, *580*, 411940. [CrossRef]
38. Manjunatha, H.; Damle, R.; Pravin, K.; Kumaraswamy, G.N. Modification in the transport and morphological properties of solid polymer electrolyte system by low-energy ion irradiation. *Ionics* **2018**, *24*, 3027–3037. [CrossRef]
39. Zhou, W.; Kou, Y.; Yuan, M.; Li, B.; Cai, H.; Li, Z.; Chen, F.; Liu, X.; Wang, G.; Chen, Q.; et al. Polymer composites filled with core@double-shell structured fillers: Effects of multiple shells on dielectric and thermal properties. *Compos. Sci. Technol.* **2019**, *181*, 107686. [CrossRef]
40. Pradhan, D.K.; Choudhary, R.N.P.; Samantaray, B.K. Studies of dielectric relaxation and AC conductivity behavior of plasticized polymer nanocomposite electrolytes. *Int. J. Electrochem. Sci.* **2008**, *3*, 597–608.
41. Aziz, S.B. Occurrence of electrical percolation threshold and observation of phase transition in chitosan $(1-x)$: AgI x $(0.05 \leq x \leq 0.2)$-based ion-conducting solid polymer composites. *Appl. Phys. A Mater. Sci. Process.* **2016**, *122*, 891–901. [CrossRef]
42. Brza, M.A.; Aziz, S.B.; Anuar, H.; Ali, F.; Abdulwahid, R.T.; Hadi, J.M. Electrochemical Impedance Spectroscopy as a Novel Approach to Investigate the Influence of Metal Complexes on Electrical Properties of Poly(vinyl alcohol) (PVA) Composites. *Int. J. Electrochem. Sci.* **2021**, *16*, 210542. [CrossRef]
43. Abiddin, J.F.B.Z.; Ahmad, A.H. Conductivity study and fourier transform infrared (FTIR) characterization of methyl cellulose solid polymer electrolyte with sodium iodide conducting ion. *AIP Conf. Proc.* **2015**, *1674*, 020026. [CrossRef]
44. Aziz, S.B.; Nofal, M.M.; Ghareeb, H.O.; Dannoun, E.M.A.; Hussen, S.A.; Hadi, J.M.; Ahmed, K.K.; Hussein, A.M. Characteristics of Poly(vinyl Alcohol) (PVA) Based Composites Integrated with Green Synthesized $Al^{3+}$-Metal Complex: Structural, Optical, and Localized Density of State Analysis. *Polymers* **2021**, *13*, 1316. [CrossRef] [PubMed]
45. Shuhaimi, N.E.A.; Teo, L.P.; Woo, H.J.; Majid, S.R.; Arof, A.K. Electrical double-layer capacitors with plasticized polymer electrolyte based on methyl cellulose. *Polym. Bull.* **2012**, *69*, 807–826. [CrossRef]
46. Priya, S.S.; Karthika, M.; Selvasekarapandian, S.; Manjuladevi, R. Preparation and characterization of polymer electrolyte based on biopolymer I-Carrageenan with magnesium nitrate. *Solid State Ionics* **2018**, *327*, 136–149. [CrossRef]
47. Samsudin, A.S.; Lai, H.M.; Isa, M.I.N. Biopolymer materials based carboxymethyl cellulose as a proton conducting biopolymer electrolyte for application in rechargeable proton battery. *Electrochim. Acta* **2014**, *129*, 1–13. [CrossRef]
48. Kasprzak, D.; Stępniak, I.; Galiński, M. Electrodes and hydrogel electrolytes based on cellulose: Fabrication and characterization as EDLC components. *J. Solid State Electrochem.* **2018**, *22*, 3035–3047. [CrossRef]
49. Aziz, S.B.; Brza, M.A.; Hamsan, H.M.; Kadir, M.F.Z.; Abdulwahid, R.T. Electrochemical characteristics of solid state double-layer capacitor constructed from proton conducting chitosan-based polymer blend electrolytes. *Polym. Bull.* **2020**, *78*, 3149–3167. [CrossRef]
50. Kadir, M.F.Z.; Arof, A.K. Application of PVA-chitosan blend polymer electrolyte membrane in electrical double layer capacitor. *Mater. Res. Innov.* **2011**, *15*, s217–s220. [CrossRef]
51. Hamsan, M.H.; Aziz, S.B.; Nofal, M.M.; Brza, M.A.; Abdulwahid, R.T.; Hadi, J.M.; Karim, W.O.; Kadir, M.F.Z. Characteristics of EDLC device fabricated from plasticized chitosan:$MgCl_2$ based polymer electrolyte. *J. Mater. Res. Technol.* **2020**, *9*, 10635–10646. [CrossRef]

52. Kadir, M.F.Z.; Salleh, N.S.; Hamsan, M.H.; Aspanut, Z.; Majid, N.A.; Shukur, M.F. Biopolymeric electrolyte based on glycerolized methyl cellulose with NH4Br as proton source and potential application in EDLC. *Ionics* **2018**, *24*, 1651–1662. [CrossRef]
53. Yusof, Y.M.; Majid, N.A.; Kasmani, R.M.; Illias, H.A.; Kadir, M.F.Z. The Effect of Plasticization on Conductivity and Other Properties of Starch/Chitosan Blend Biopolymer Electrolyte Incorporated with Ammonium Iodide. *Mol. Cryst. Liq. Cryst.* **2014**, *603*, 73–88. [CrossRef]
54. Brza, M.; Aziz, S.B.; Saeed, S.R.; Hamsan, M.H.; Majid, S.R.; Abdulwahid, R.T.; Kadir, M.F.Z.; Abdullah, R.M. Energy storage behavior of lithium-ion conducting poly(Vinyl alcohol) (pva): Chitosan(cs)-based polymer blend electrolyte membranes: Preparation, equivalent circuit modeling, ion transport parameters, and dielectric properties. *Membranes* **2020**, *10*, 381. [CrossRef] [PubMed]
55. Bandaranayake, C.M.; Weerasinghe, W.A.D.S.S.; Vidanapathirana, K.P.; Perera, K.S. A Cyclic Voltammetry study of a gel polymer electrolyte based redox-capacitor. *Sri Lankan J. Phys.* **2016**, *16*, 19–27. [CrossRef]
56. Winie, T.; Jamal, A.; Saaid, F.I.; Tseng, T.Y. Hexanoyl chitosan/ENR25 blend polymer electrolyte system for electrical double layer capacitor. *Polym. Adv. Technol.* **2019**, *30*, 726–735. [CrossRef]
57. Aziz, S.B.; Hamsan, M.H.; Karim, W.O.; Marif, A.S.; Abdulwahid, R.T.; Kadir, M.F.Z.; Brza, M.A. Study of impedance and solid-state double-layer capacitor behavior of proton ($H^+$)-conducting polymer blend electrolyte-based CS:PS polymers. *Ionics* **2020**, *26*, 4635–4649. [CrossRef]
58. Hadi, J.M.; Aziz, S.B.; Saeed, S.R.; Brza, M.A.; Abdulwahid, R.T.; Hamsan, M.H.; Abdullah, R.M.; Kadir, M.F.Z.; Muzakir, S.K. Investigation of ion transport parameters and electrochemical performance of plasticized biocompatible chitosan-based proton conducting polymer composite electrolytes. *Membranes* **2020**, *10*, 363. [CrossRef]
59. Kumar, Y.; Pandey, G.P.; Hashmi, S.A. Gel polymer electrolyte based electrical double layer capacitors: Comparative study with multiwalled carbon nanotubes and activated carbon electrodes. *J. Phys. Chem. C* **2012**, *116*, 26118–26127. [CrossRef]
60. Muchakayala, R.; Song, S.; Wang, J.; Fan, Y.; Bengeppagari, M.; Chen, J.; Tan, M. Development and supercapacitor application of ionic liquid-incorporated gel polymer electrolyte films. *J. Ind. Eng. Chem.* **2018**, *59*, 79–89. [CrossRef]
61. Kamarudin, K.H.; Hassan, M.; Isa, M.I.N. Lightweight and flexible solid-state EDLC based on optimized CMC-$NH_4NO_3$ solid bio-polymer electrolyte. *ASM Sci. J.* **2018**, *11*, 29–36.
62. Aziz, S.B.; Hamsan, M.H.; Brza, M.A.; Kadir, M.F.Z.; Muzakir, S.K.; Abdulwahid, R.T. Effect of glycerol on EDLC characteristics of chitosan:methylcellulose polymer blend electrolytes. *J. Mater. Res. Technol.* **2020**, *9*, 8355–8366. [CrossRef]
63. Yang, Z.; Peng, H.; Wang, W.; Liu, T. Crystallization behavior of poly(ε-caprolactone)/layered double hydroxide nanocomposites. *J. Appl. Polym. Sci.* **2010**, *116*, 2658–2667. [CrossRef]
64. Shukur, M.F.; Hamsan, M.H.; Kadir, M.F.Z. Investigation of plasticized ionic conductor based on chitosan and ammonium bromide for EDLC application. *Mater. Today Proc.* **2019**, *17*, 490–498. [CrossRef]
65. Yusof, Y.M.; Shukur, M.F.; Hamsan, M.H.; Jumbri, K.; Kadir, M.F.Z. Plasticized solid polymer electrolyte based on natural polymer blend incorporated with lithium perchlorate for electrical double-layer capacitor fabrication. *Ionics* **2019**, *25*, 5473–5484. [CrossRef]

*Article*

# The Influence of the Electrodeposition Parameters on the Properties of Mn-Co-Based Nanofilms as Anode Materials for Alkaline Electrolysers

**Karolina Cysewska [1,*], Maria Krystyna Rybarczyk [2], Grzegorz Cempura [3], Jakub Karczewski [4], Marcin Łapiński [4], Piotr Jasinski [1] and Sebastian Molin [1]**

1 Faculty of Electronics, Telecommunications and Informatics, Gdansk University of Technology, ul. Narutowicza 11/12, 80-233 Gdansk, Poland; piotr.jasinski@pg.edu.pl (P.J.); sebastian.molin@pg.edu.pl (S.M.)
2 Department of Process Engineering and Chemical Technology, Chemical Faculty, Gdansk University of Technology, ul. Narutowicza 11/12, 80–233 Gdansk, Poland; maria.rybarczyk@pg.edu.pl
3 International Centre of Electron Microscopy for Materials Science, Faculty of Metals Engineering and Industrial Computer Science, AGH University of Science and Technology, ul. A. Mickiewicza 30, 30-059 Krakow, Poland; cempura@agh.edu.pl
4 Faculty of Applied Physics and Mathematics, Gdansk University of Technology, ul. Narutowicza 11/12, 80–233 Gdansk, Poland; jakkarcz@pg.edu.pl (J.K.); marcin.lapinski@pg.edu.pl (M.Ł.)
* Correspondence: karolina.cysewska@pg.edu.pl

Received: 21 May 2020; Accepted: 10 June 2020; Published: 11 June 2020

**Abstract:** In this work, the influence of the synthesis conditions on the structure, morphology, and electrocatalytic performance for the oxygen evolution reaction (OER) of Mn-Co-based films is studied. For this purpose, Mn-Co nanofilm is electrochemically synthesised in a one-step process on nickel foam in the presence of metal nitrates without any additives. The possible mechanism of the synthesis is proposed. The morphology and structure of the catalysts are studied by various techniques including scanning electron microscopy, X-ray diffraction, X-ray photoelectron spectroscopy, and transmission electron microscopy. The analyses show that the as-deposited catalysts consist mainly of oxides/hydroxides and/or (oxy)hydroxides based on $Mn^{2+}$, $Co^{2+}$, and $Co^{3+}$. The alkaline post-treatment of the film results in the formation of Mn-Co (oxy)hydroxides and crystalline $Co(OH)_2$ with a β-phase hexagonal platelet-like shape structure, indicating a layered double hydroxide structure, desirable for the OER. Electrochemical studies show that the catalytic performance of Mn-Co was dependent on the concentration of Mn versus Co in the synthesis solution and on the deposition charge. The optimised Mn-Co/Ni foam is characterised by a specific surface area of 10.5 $m^2 \cdot g^{-1}$, a pore volume of 0.0042 $cm^3 \cdot g^{-1}$, and high electrochemical stability with an overpotential deviation around 330–340 mV at 10 $mA \cdot cm^{-2}{}_{geo}$ for 70 h.

**Keywords:** alkaline electrolyser; electrocatalyst; electrodeposition; energy material; nanofilm; nickel foam; oxygen evolution reaction

## 1. Introduction

The oxygen evolution reaction (OER) has become the main limitation to the efficiency of the water splitting process [1,2]. In order to make the reaction more robust, novel anode catalysts, which would lower the overpotential needed to drive the reaction, are required [3]. Recently, both $RuO_2$ and $IrO_2$ have become highly active, benchmark catalysts as anode materials for the OER with an overpotential typically close to 350 mV at 10 $mA \cdot cm^{-2}$ [4]. However, their limited sources, high cost, and inferior stability at higher anodic potentials do not allow for large-scale usage [5,6]. Therefore, recently, there has been a huge effort to fabricate nanostructured oxide/hydroxide electrocatalysts based on earth-abundant elements [7–9]. Because of their eco-friendly properties and low cost, they have become interesting materials for different energy applications including batteries [10], supercapacitors [11], fuel cells [12], and alkaline water electrolyzers [13,14].

The traditional method for the synthesis of oxide/hydroxide films is based on the solid-state approach, which includes grinding and firing a mixture of certain metal oxides, nitrates, or carbonates [15,16]. Other possible chemical methods are sol-gel [17,18], combustion [19,20], and hydro/solvothermal routes [21,22]. However, all of the above-mentioned synthesis methods typically lead to the formation of pure catalyst powder, which requires further processing to produce an ink containing other additives such as a binder and conductive carbon powder [15,23]. This, in turn, may result in very poor stability of the OER electrode [24,25]. One of the promising, alternative synthesis techniques is electrodeposition. It allows for the formation of the catalyst directly on the conductive substrate without any additives [26,27]. Moreover, the properties of the deposited films can be easily tailored by changing the synthesis conditions such as the electrolyte concentration, type of solvent, and/or type of electrodeposition (potentiostatic/galvanostatic) [28]. By growing the catalysts directly on the substrate, their adherence and integrity, based on chemical bonding, should be superior to those of ink-based catalysts.

In the literature, there are several attempts to electrodeposit oxides/hydroxides based on different transition metals on conductive substrates [26,29–32]. Vigil et al. [33] reported the electrodeposition process for manganese oxide ($MnO_x$) together with the conducting polymer 3,4-ethylenedioxytiophene (PEDOT) on a glassy carbon electrode for energy storage and conversion devices. Other studies proposed the electrochemical deposition of $Mn_{1.5}Co_{1.5}O_4$ on Ni foam [31], Mn oxide rods on a gold/silica substrate [32], and Mn/Co or Ni/Co hydroxides on stainless steel [13,34] for supercapacitors.

Different kinds of electrodeposited oxides/hydroxides have also been studied as possible anode materials for the OER in alkaline environments [24,35–38]. A. Ramirez [35] studied the OER activity of differently electrodeposited manganese oxides such as $MnO_x$, $Mn_2O_3$, and $Mn_3O_4$ on F:$SnO_2$/glass. F. Yan reported the electrocatalytic properties of $MnO_2$ on carbon cloth with an overpotential of 424 mV at 10 mA·cm$^{-2}$ [24]. A lower OER overpotential of the electrode compared to a bare substrate was obtained for cobalt oxide synthesised on platinum, nickel, and iron [36,39,40]. P. Liu et al. [41] fabricated electrochemically etched α-cobalt hydroxide for the OER, characterising by an overpotential of 320 mV at 10 mA·cm$^{-2}$.

Recently, significant effort has been made to synthesise mixed oxides/hydroxides based on Co [38], Mn [34], Ni, and Fe [37] due to their high electrocatalytic activity coming from their abundant defects and fast redox reactions [31]. It was shown that the OER overpotential could be decreased by introducing cobalt into oxides in different valence states [42]. It is also interesting to incorporate manganese into mixed oxides. It is cheap and environmentally friendly, and provides the possibility of several redox reactions due to its multi-valance state [29]. F. Yan presented the electrocatalytic properties of MnCo-layered double hydroxide (LDH) on carbon cloth, which exhibited an overpotential of 258 mV at 10 mA·cm$^{-2}$ [24]. F.J. Perez-Alonso et al. [37] reported Ni/Fe-based oxides with different compositions on Ni foam and stainless steel. Zn/Co-based spinels such as $Zn_xCo_3$-x$O_4$ on gold [38] and $ZnCo_2O_4$ on platinum [43] have also been developed for the OER. Their overpotentials were determined to be 330 mV and 390 mV at 10 mA·cm$^{-2}$, respectively. Bao et al. [44] introduced Mn-Co based LDH-graphene composite, which revealed an OER overpotential of 330 mV at 10 mA·cm$^{-2}$ and showed superior bifunctional water splitting activity. In another work, (Co, Ni)Mn-LDH nanosheets were fabricated on multi-wall carbon nanotubes for efficient OER. The material exhibited an OER overpotential of 300 mV at 10 mA·cm$^{-2}$ and revealed excellent long-term electrocatalytic stability [45]. The previous work of Lankauf et al. [46] presented $Mn_xCo_{3-x}O_4$ deposited on nickel foam with an OER overpotential of 327 mV at 10 mA·cm$^{-2}$ and only slight material degradation after exposure to an alkaline environment for 25 h. Recently, Mn/Co-based film materials have also become promising catalysts in sealed-oxygen batteries for triggering oxygen-related anionic redox activity [47].

It is well known that the electrocatalytic properties of the material strongly depend on the type of the substrate. Studies have shown that the most suitable active metal for an alkaline electrolyser is nickel [37,48]. It is relatively cheap and chemically stable. The porous nature of nickel provides a much higher electroactive surface area compared to planar electrodes [48]. This in turn supports the ion/electron conduction between the catalyst and electrolyte, which results in higher OER activity of the electrode. It should be noted that there are almost no reports on electrodeposited Mn-Co-based oxides/hydroxides on a nickel substrate for alkaline electrolysers. Additionally, most of the performed studies were focused on investigating the properties of the catalysts after thermal treatment (spinels). There is a lack of literature reports on the structural and OER activity of as-deposited different forms of Mn/Co oxides/hydroxides without any post-treatment.

In this work, electrodeposited Mn-Co-based nanofilms on Ni foam were studied as potential catalysts for the oxygen evolution reaction. The catalysts were synthesised in a one-step process in an aqueous solution of manganese and cobalt nitrates without any additives. The possible mechanism of such a synthesis was proposed. The films were studied with regard to their structural and electrocatalytic properties for the OER. Mainly, the influence of the concentration ratio of Mn/Co nitrates used in a synthesis solution and deposition charge were investigated.

## 2. Materials and Methods

### 2.1. Chemicals and Materials

The chemicals used in this work were manganese (II) nitrate tetrahydrate ($Mn(NO_3)_2{\cdot}4H_2O$) (99%, Panreac), cobalt (II) nitrate hexahydrate ($Co(NO_3)_2{\cdot}6H_2O$) (98%, Sigma Aldrich, Saint Louis, MO, USA), and potassium hydroxide (KOH) (Stanlab, Lublin, Poland). The substrate was a pure nickel foam (~96% porosity, 110 pores per inch) or nickel foil. The solutions were prepared with distilled water (Millipore Elix Essential 3, Millipore corporation, Billerica, MA, U.S.A, >12 MΩ cm).

Before each electrodeposition process, the substrate was ultrasonically cleaned in distilled water and acetone, respectively for 5 min each, and dried in the air.

### 2.2. Preparation of Electrocatalysts

Mn-Co-based films were electrochemically synthesised in a one-step process at −1.1 V vs. Ag/AgCl in an aqueous solution of differently concentrated $Mn(NO_3)_2{\cdot}4H_2O$ and $Co(NO_3)_2{\cdot}6H_2O$ (≈pH 3) with the deposition time limited by charges of 60, 120, and 200 mC at 25 °C. The concentration ratios of $Mn(NO_3)_2{\cdot}4H_2O$ to $Co(NO_3)_2{\cdot}6H_2O$ were 2 mM to 4 mM, 2 mM to 6 mM, 2 mM to 8 mM, 4 mM to 2 mM, and 0 mM to 4 mM.

The experiments were performed in a one-compartment water-jacketed, three-electrode cell controlled by a VersaSTAT 4 potentiostat. The working electrode was nickel foam or nickel foil (0.5 cm × 0.5 cm) with a working area of 0.25 $cm^2$. An Ag/AgCl in 3M KCl was the reference electrode (Hydromet) and a coiled platinum wire was the counter electrode.

### 2.3. Electrochemical Measurements

The electrochemical measurements were conducted in the same setup as the electrodeposition process (see Section 2.2) in an aqueous solution of 1 M KOH with a measured pH ≈ 13.9 at 25 °C. The temperature was controlled by a JULABO F12 thermostat. Before each measurement, the electrolyte was continuously purged with Ar for 20 min. The linear scan voltammetry (LSV) data were recorded from 0.1 to 1 V vs. Ag/AgCl with a scan rate of 5 $mV{\cdot}s^{-1}$. The electrochemical impedance spectra (EIS) were acquired in the frequency range of 10 kHz–0.1 Hz at 0.7 V vs. Ag/AgCl with an amplitude of 10 mV. The EIS data were fitted with the Zview software. The double layer capacitance ($C_{dl}$) was determined based on the cyclic voltammetry (CV) performed by sweeping the potential across the non-faradaic region from 0.15 V to 0.25 V vs. Ag/AgCl at different scan rates (10, 20, 40, 60, 80, 100 $mV{\cdot}s^{-1}$) in an aqueous solution of 1 M KOH (Figure S1a,c,e, Supplementary content). The CV allowed for the determination of $C_{dl}$ based on the following equation $C_{dl} = i_{dl}\cdot(2\upsilon)^{-1} = (i_a - i_c)\cdot(2\upsilon)^{-1}$, where $i_{dl}$ is the double-layer current density; $i_a$ and $i_c$ are the anodic and cathodic current densities, respectively; and $\upsilon$ is the scan rate. Thus, plotting half of the double-layer current density as a function of the scan rates yielded straight lines with slopes equal to the double-layer capacitance (Figure S1b,d,f, Supplementary content). The electrochemical surface area (ECSA) of the samples was determined according to the equation ECSA = $C_{dl}\cdot A\cdot {C_{spec}}^{-1}$, where A is the geometric surface area of the sample and $C_{spec}$ is the specific capacitance of 0.040 $mF\cdot {cm^{-2}}_{geo}$, which is a typical value reported for a metal electrode in an aqueous alkaline solution [49]. The stability test was performed at 10 $mA\cdot cm^{-2}$ for approximately 70 h. Before each electrochemical measurement, CV of the studied material was carried out in order to obtain an activated and stable system. For this purpose, the CV was carried out from 0.1 to 0.6 V vs. Ag/AgCl with a scan rate of 100 $mV\cdot s^{-1}$ for at least 20 cycles.

All of the potentials were calibrated to the reversible hydrogen electrode potentials (vs. RHE) according to the equation $E_{vs.\,RHE} = E_{vs.Ag/AgCl} + {E^0}_{Ag/AgCl\ in\ 3MKCl} + 0.059$ pH. Unless otherwise stated, the values of all potentials were iR-corrected to remove the effect of the solution resistance according to the equation $E_{iR\text{-}corrected} = E_{applied} - iR_{un}$, where i is the current and $R_{un}$ is uncompensated ohmic electrolyte resistance in an Ar-saturated 1 M KOH solution. The overpotential ($\eta$) for the oxygen evolution reaction was calculated by the following equation: $\eta$ = E (10 $mA\cdot cm^{-2}$) − 1.23 V (vs. RHE) [50].

### 2.4. Material Characterisation

The morphology and elemental analysis of the catalysts synthesised under different conditions were investigated using a FEI QUANTA FEG 250 scanning electron microscope (SEM, Thermofisher (FEI), Waltham, MA, USA) with energy dispersive X-ray (EDX) analysis and with a FEI Titan G2-300 transmission electron microscope (TEM, Thermofisher (FEI), Waltham, MA, USA). For the TEM, the Mn-Co film was scratched off the nickel substrate.

X-ray photoemission spectroscopy (XPS, Omicron NanoTechnology, Taunusstein, Germany) measurements were carried out with Omicron NanoTechnology ultra-high vacuum equipment. The hemispherical spectrophotometer was equipped with a 128-channel collector. The XPS measurements were performed at room temperature at a pressure below $1.1 \times 10^{-8}$ mBar. The photoelectrons were excited by an Mg-K$\alpha$ X-Ray source. The X-ray anode was operated at 15 keV and 300 W. The results were corrected using the C1s peak (285.0 eV). The XPS spectra were analysed with the Casa-XPS software using a Shirley background subtraction and Gaussian–Lorentzian (GL30) curve as a fitting algorithm. XPS spectra were fitted with residual standard deviations (STDs) lower than 0.8 and 1.1 for the Co2p and Mn2p lines, respectively.

X-ray diffraction (XRD, Bruker, Billerica, MA, USA) measurements were conducted on a Bruker D2 Phaser 2nd generation diffractometer with CuK$\alpha$ radiation ($\lambda$ = 1.5404 Å). Data were collected from $2\Theta = 10°$ to 90° with a step size of 0.01° at room temperature. XRD measurement was performed for optimised Mn-Co on nickel foil deposited from the solution of Mn/Co 2 mM/8 mM for 200 mC (before and after alkaline treatment in 1 M KOH).

The specific surface area of the samples was also characterised by a Brunauer–Emmett–Teller (BET, Micromeritics Instruments Corporation, Norcross, GA, USA) Micromeritics Gemini V apparatus (model 2365). For each measurement, the sample weight was in the range of 0.01–0.02 g. The degassing temperature reached 120 °C. The contact angle of the electrodes was determined by an OCA15 goniometer. The liquid used was distilled water. The non-local density functional theory (DFT) model was applied in order to assess the pore characteristics. The pore volume was calculated using the normal liquid density of the adsorbate [51].

## 3. Results and Discussion

### 3.1. Electrochemical Formation and Morphology of Mn-Co-Based Oxides/Hydroxides on Ni Foam

All manganese–cobalt-based films (Mn-Co) were electrochemically synthesised in a one-step process in an aqueous solution of differently concentrated $Mn(NO_3)_2{\cdot}4H_2O$ (Mn) and $Co(NO_3)_2{\cdot}6H_2O$ (Co) without any additives. The synthesis graphs recorded during the electrodeposition are presented in Figure 1a.

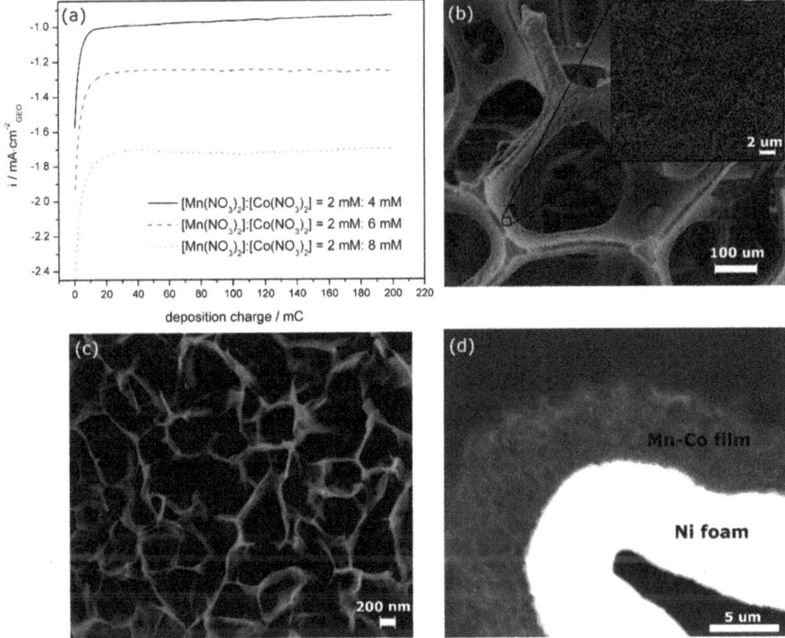


**Figure 1.** (**a**) Synthesis graphs recorded during the electrochemical synthesis of Mn/Co oxide/hydroxides at −1.1 V vs. Ag/AgCl in an aqueous solution of differently concentrated $Mn(NO_3)_2{\cdot}4H_2O$ and $Co(NO_3)_2{\cdot}6H_2O$ with the deposition time limited by a charge of 200 mC; (**b–d**) SEM images of Mn-Co film deposited on nickel foam in a solution of Mn/Co (2 mM/8 mM) for 200 mC with different magnifications (**b**,**c**—surfaces, **d**—polished cross-section).

At the beginning of the deposition, the cathodic current density decreases very fast, indicating the growth of a new phase on the nickel substrate [13]. After a certain time, when a total charge of ~10 mC is reached for all of the samples, the current density becomes stable, due to the steady state formation of Mn-Co deposits. The synthesis trend is similar for all studied cases. A higher value of the cathodic current density is found for the higher concentrations of $Co(NO_3)_2{\cdot}6H_2O$ in the synthesis solution. Moreover, the current density during the steady-state deposition increases linearly with a higher cobalt concentration in the solution. For comparison, Co and Mn-Co films were deposited on nickel foam in the presence of only 4 mM $Co(NO_3)_2{\cdot}6H_2O$ (Mn/Co 0 mM/4 mM) or in the presence of a higher content of manganese compared to cobalt, i.e., 4 mM $Mn(NO_3)_2{\cdot}4H_2O$/2 mM $Co(NO_3)_2{\cdot}6H_2O$ (Mn/Co 4 mM/2 mM). The synthesis curves recorded for such cases reveal that the deposition process in the presence of cobalt alone proceeds much more slowly (lower cathodic current density) compared to the synthesis in the presence of both Mn and Co (Figure S2, Supplementary content). The maximum cathodic current density for Mn/Co 0 mM/4 mM and Mn/Co 2 mM/4 mM was determined to be approximately $-0.22$ mA·cm$^{-2}{}_{geo}$ and $-1$ mA·cm$^{-2}{}_{geo}$, respectively. Moreover, the current density achieved during the deposition of Mn-Co in the presence of a higher concentration of manganese was determined to be −0.33 mA·cm$^{-2}$$_{geo}$, which is much lower compared to Mn-Co with a higher content of cobalt (Figure S2, Supplementary content). These all indicate that the nucleation rate is the fastest in the presence of both $Mn(NO_3)_2{\cdot}4H_2O$ and $Co(NO_3)_2{\cdot}6H_2O$ with a higher content of cobalt in the ratio.

The as-deposited Mn-Co films were characterised by a brownish/gold and green colour, indicating the presence of manganese and cobalt oxides/hydroxides, respectively. Here, it should also be noted that soaking the as-prepared samples in an aqueous solution of 1 M KOH resulted in a changing of the catalyst's colour from brownish/gold-green to black, suggesting a change in the catalyst's structure (Figure S3, Supplementary content). Because of that, it was important to also study the properties of the films after alkaline treatment in 1 M KOH, which will be related to the properties during the measurements.

The morphology of the differently prepared samples was studied by scanning electron microscopy (SEM). Figure 1b–d present SEM images of Mn-Co films on nickel foam deposited in an aqueous solution of Mn/Co 2 mM/8 mM with the deposition time limited by a charge of 200 mC. The SEM analysis confirmed the successful deposition of the Mn-Co-based film on the porous nickel foam (Figure 1b,c). The structure of the as-deposited Mn-Co films was characterised by interconnected nanoflakes uniformly deposited on the nickel substrate, forming a porous interconnected 3D network. The thin nanoflake structure was also observed for a chemically synthesised Mn-Co catalyst [44]. Such a morphology might support the fast transport of hydroxide ions (OH-) due to the easily accessible open spaces [30]. This, in turn, should ensure high structural stability in the OER [24]. There was no significant change in the surface topography between the Mn/Co films synthesised in the aqueous solutions of Mn/Co 2 mM/4 mM, 2 mM/6 mM, and 2 mM/8 mM (Figure S4, Supplementary content). A similar nanosheet structure was obtained when the substrate was nickel foil (Figure S5a, Supplementary content).

The structure of the catalyst after synthesis in the presence of only 4 mM $Co(NO_3)_2{\cdot}6H_2O$ revealed mainly a nanosheet-like structure, which in some places agglomerated (Figure S6a, Supplementary content). The film consisted only of the cobalt compounds, which was proven by the EDX elemental analysis (not shown here). A totally different morphology was observed for the Mn/Co film deposited in the presence of a higher concentration of Mn compared to that in cobalt Mn/Co 4 mM/2 mM (Figure S6b, Supplementary content). Here, the structure was characterised by randomly distributed and much smaller nanosheets with star-like structures with an atomic ratio of Mn to Co of 3.7:1. The results indicate that the nanosheet structure of the as-deposited Mn/Co film prepared in the solutions of Mn/Co 2 mM/4 mM, 2 mM/6 mM, and 2 mM/8 mM is mainly determined by the $Co(NO_3)_2{\cdot}6H_2O$.

The thickness of the deposited catalysts was determined based on the cross-sectional SEM images and varied from 2 to 7 μm for Mn-Co 2 mM/8 mM electrodeposited with a charge of 200 mC on nickel foam. Figure 1d presents a cross-sectional SEM image of the thickest part of the film obtained on nickel foam. For comparison, the same Mn-Co film was electrodeposited on nickel foil. The cross-sectional SEM image shows that the catalyst was homogenously deposited on the foil (Figure S5b, Supplementary content). The thickness of the film was determined to be approximately 2 μm. The irregular deposition of the catalyst on the nickel foam might be either due to its highly porous nature or the chosen deposition method. In the case of such a substrate, the efficiency of the electrodeposition might differ due to the difficulties of introducing the synthesis solution into the cavities of the foam, which is not the case when the planar electrode is used. The thickness of the Mn-Co film was virtually the same for the film synthesised in the presence of differently concentrated Mn/Co nitrates in the solution.

The mass of the catalyst synthesised with the limited charge of 200 mC was determined to be around 50 μg for each studied case (based on a weight gain measurement on a microbalance). The theoretical mass was estimated based on Faraday's law and was determined to be around 186 μg, assuming that the as-deposited film consists mainly of an LDH structure of CoOOH and MnOOH, with a total molar mass (M) of 179.87 g·mol$^{-1}$, an electron loss (z) of 2, and a deposition charge of 200 mC. The differences in the experimental and theoretical mass are mainly due to the assumptions, which can greatly influence the theoretical calculations. It also indicates that the deposition process did not proceed with 100% efficiency.

### 3.2. XRD, XPS, and TEM Analysis of Mn-Co-Based Nanofilm

The structure of the as-deposited Mn-Co film (2 mM/8 mM) on nickel foil and after alkaline treatment in 1 M KOH was analysed by X-ray diffraction (XRD), and the results are presented in Figure 2a.

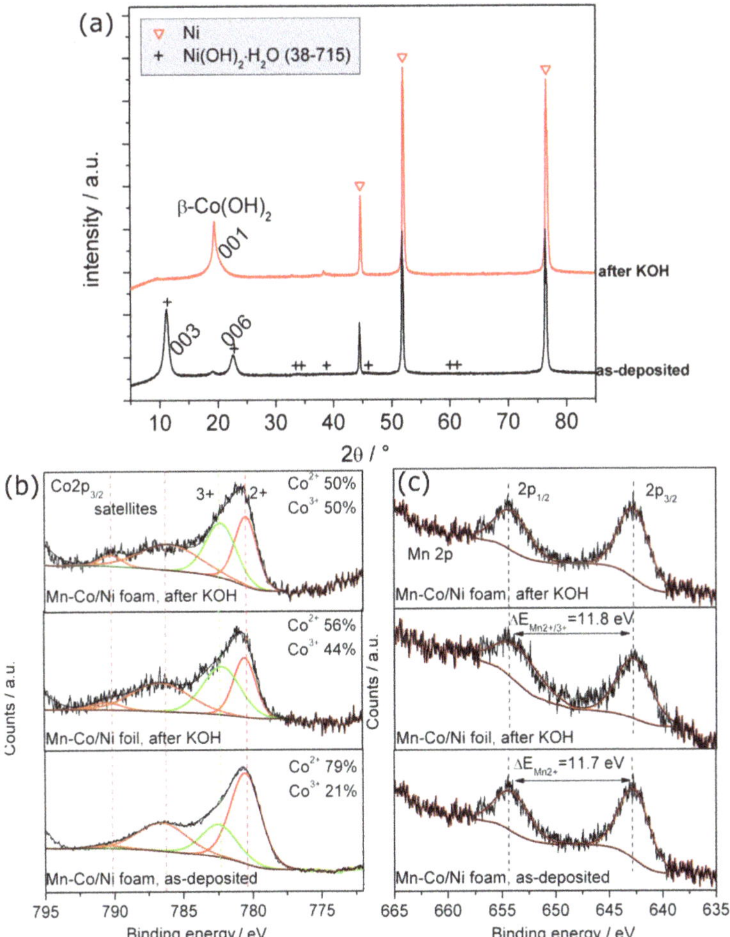


**Figure 2.** (**a**) XRD patterns of Mn-Co deposited on nickel before (red, triangle), and after (black, cross) alkaline treatment. The most intense diffraction peaks are assigned to Ni foil. XPS scans of (**b**) Co and (**c**) Mn of as-deposited and after-alkaline-treatment Mn-Co film. For XRD and XPS, the film was synthesised in a solution of Mn/Co 2 mM/8 mM for 200 mC on nickel foam/foil.

The peaks at 2θ of around 45°, 52°, and 76° associated with the presence of nickel are observed in each studied case, which is in agreement with the literature [52]. The rest of the peaks observed for the as-deposited Mn-Co film can be attributed to nickel hydroxide in accordance with the JCDS database (38–715). The presence of both planes (003) and (006) at 2θ of 10.8° and 22.5°, respectively, was related to the presence of α-phase cobalt hydroxide $Co(OH)_2$ [34,53,54], which in the literature was assigned to the typical pattern of layered double hydroxides (LDH) [52,55]. Soaking the as-deposited Mn-Co film in 1 M KOH results in the formation of crystalline β-phase cobalt (II) hydroxide $Co(OH)_2$, which can be clearly seen in the XRD spectra as a peak at around 19.3° [54]. Such a peak can also be observed for the as-deposited sample but with a much lower intensity. Comparing the results with the Pourbaix diagram, it can be noted that in a solution of alkaline pH, the formation of $Co(OH)_2$ is expected.

The XPS analysis (Figure 2b,c) reveals the presence of the manganese and cobalt elements in the deposited film. In the Co 2p spectra (Figure 2b), two kinds of cobalt species are observed for each studied case, $Co^{2+}$ and $Co^{3+}$, at binding energies of 780.5 eV and 782.5 eV, respectively [52,56]. The Mn 2p spectra (Figure 2c) are split into Mn $2p_{1/2}$ and Mn $2p_{3/2}$, which are located at 653.5 eV and 641.8 eV, respectively, for the as-deposited film, indicating the presence mainly of $Mn^{2+}$ [57,58]. After the alkaline post-treatment of the sample in KOH solution, the energy changes at around 11.8 eV, indicating the presence of both $Mn^{2+}$ and possibly $Mn^{3+}$ [57]. The percentage contents of certain elements in the as-deposited film and the film after alkaline treatment are presented in Table 1.

**Table 1.** The percentage content of certain elements in the as-deposited Mn-Co film and the film after alkaline treatment determined based on the XPS (surface concentration, error ≤ 5%).

| Element | As-Deposited Film | Film after Alkaline Treatment |
|---|---|---|
| $Co^{2+}$ | 79% | 50% |
| $Co^{3+}$ | 21% | 50% |
| $Mn^{2+}$ | 100% | Difficult to determine |
| $Mn^{3+}$ | 0% | Difficult to determine |

The analysis shows that the content of $Co^{2+}$ to $Co^{3+}$ changes in the Mn-Co film after alkaline treatment, indicating the oxidation of the $Co^{2+}$ to $Co^{3+}$. The post-treated Mn-Co/nickel reveals the possible presence of both $Mn^{2+}$ and $Mn^{3+}$, which indicates an oxidation of manganese due to the alkaline environment. It should be noted that in the case of manganese, the differences in binding energy related to certain valence states are very small, which cannot be interpreted with 100% accuracy based only on the XPS analysis. This is also the reason why in the literature, the same energy can be associated with different valence states of manganese in the film [52,59].

The content of the $Co^{2+}$, $Co^{3+}$, and $Mn^{2+/3+}$ elements in Mn-Co film after alkaline treatment deposited either on nickel foam or nickel foil is virtually the same, indicating the good reproducibility of both the applied synthesis method and the post-treatment process in 1 M KOH.

The morphology and microstructure of the Mn-Co film on nickel foam before (Figure 3a,b) and after (Figure 3c–e) alkaline treatment in 1 M KOH were also studied by transmission emission microscopy (TEM).

As can be seen from Figure 3a,b, the as-deposited film is characterised by the nanosheet, a mainly amorphous structure, which partially crystallised due to the formation of $Co(OH)_2$ (Figure 2a). The crystallinity of the deposits increases upon alkaline treatment in KOH. The hexagonal platelet-like shape structure appeared after soaking the material in KOH solution (Figure 3c), which is a typical morphology for layered double-hydroxides [55]. The crystalline structure of the post-treated film is also proven by the Selected Area Diffraction (SAED) analysis (Figure 3d,e). The analysis of the experimental SAED pattern confirms the presence of crystalline particles of cobalt manganese (IV/VI) oxide (0.25/1.75/4). The results obtained by TEM are in agreement with the previous SEM, XPS, and XRD analyses (Figures 1 and 2). The TEM elemental mapping images of the post-treated Mn-Co film confirms the successful synthesis of the Mn-Co-based film (Figure 3f–i). The analysis demonstrates that the film is composed of uniformly distributed cobalt (47.4 at.%), manganese (6.6 at.%), and oxygen (46 at.%). Thus, the atomic ratio of Mn to Co for the film synthesised in an aqueous solution of Mn/Co 2 mM/8 mM (200 mC) equals around 1:7.2. The atomic ratios of Mn to Co for 2 mM/4 mM and 2 mM/6 mM Mn/Co in the synthesis solution were determined to be 1:3.1 and 1:9.5, respectively. This indicates that the atomic ratio of Mn to Co in the film was not linearly related to the concentration of both the cobalt and manganese nitrates used in the synthesis solution.

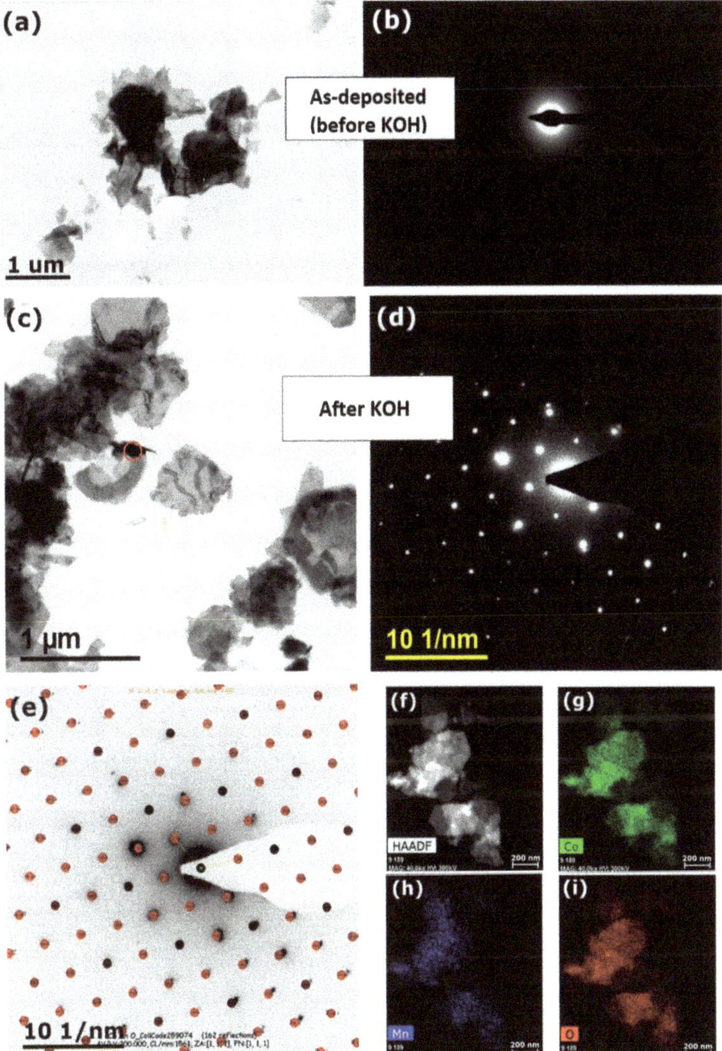


**Figure 3.** (**a**) Bright Field (BF) TEM image, and (**b**) corresponding experimental Selected Area Diffraction (SAED) pattern of Mn-Co film before alkaline treatment; (**c**) BF-TEM image, area of SAED pattern is marked with red circle, and (**d**) corresponding experimental SAED pattern from the area, marked on a BF-TEM image with a red circle, of Mn-Co film after alkaline treatment; (**e**) Experimental SAED pattern of area marked with a red circle, superimposed with simulated, theoretical diffractogram from cobalt manganese (IV/VI) oxide (0.25/1.75/4), (259073-ICSD) Zone Axis [111]; (**f–i**) TEM elemental mapping images of (**g**) cobalt, (**h**) manganese, and (**i**) oxygen of the Mn-Co film. Mn-Co film for TEM analysis was synthesised in an aqueous solution of Mn/Co 2 mM/8 mM (200 mC).

The specific surface areas of the films were analysed based on the Brunauer–Emmett–Teller (BET) method. Figure 4 presents the evolution of the BET-specific surface area (a) and pore volume (b) determined for bare nickel foam and nickel foam coated with a Mn-Co film synthesised under different conditions as a function of the deposition charge (Qd).

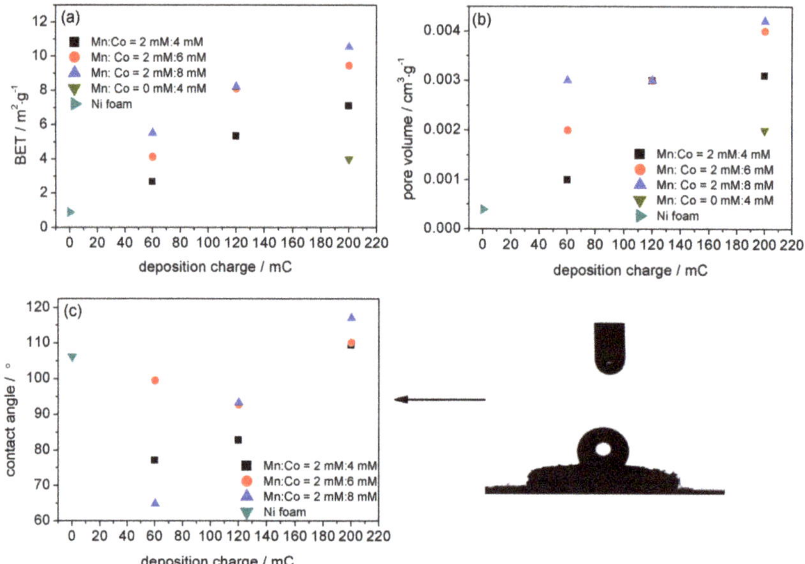


**Figure 4.** Evolution of (**a**) Brunauer–Emmett–Teller (BET) specific surface area, (**b**) pore volume, and (**c**) contact angle with an example of an optical microscopy image taken during the measurement of Mn/Co 2 mM/8 mM for bare nickel foam and nickel foam coated with Mn-Co film synthesised under different conditions as a function of the deposition charge.

The graph clearly shows that the BET surface area increases gradually with both a higher content of cobalt with respect to manganese nitrates in the synthesis solution and with the deposition charge. Even though the higher concentration of cobalt versus manganese nitrates in the synthesis solution did not induce any significant changes in the film morphology in the SEM images (Figure S4, Supplementary content), it strongly influenced the level of the specific surface area. Probably, the higher content of cobalt induced the formation of additional nanosheets, forming a percolation network with an open porous structure, which resulted in the higher specific surface area [29]. A higher surface area was also observed in the case of the higher deposition charge. This can be explained by the fact that a higher deposition charge for the same geometric area of the electrode results in a thicker deposited film [60]. This in turn, led to a higher specific surface area, which could be seen for each studied case in Figure 4a.

The evolution of the BET surface area also shows that each of the Mn-Co films prepared in solutions of Mn/Co 2 mM/4 mM, 2 mM/6 mM, and 2 mM/8 mM reveal a higher specific surface area compared to bare nickel foam. The highest surface area of 10.5 $m^2 \cdot g^{-1}$ was obtained for the Mn-Co film prepared in the solution of Mn/Co 2 mM/8 mM for 200 mC, which is almost 12 times higher in comparison to that of bare nickel foam.

The specific surface area of the film on nickel foam synthesised only in the presence of 4 mM cobalt nitrate with 200 mC was determined to be 4 $m^2 \cdot g^{-1}$, which is 2.6 times lower than that of the film synthesised in the presence of both manganese and cobalt compounds.

The evolution of the corresponding pore volume for bare nickel foam and nickel foam coated with a Mn-Co film synthesised under different conditions as a function of the deposition charge is presented in Figure 4b. The trend is very similar to the trends obtained from the BET surface area, i.e., the pore volume of the film on nickel foam increases with an increasing deposition charge and with a higher content of cobalt in the Mn/Co concentration ratio in the solution. A higher pore volume indicates a more porous catalyst. The Ni foam coated with the Mn-Co film reveals a higher pore volume compared to bare nickel. Mn-Co-based nickel foam synthesised in the presence of Mn/Co 2 mM/8 mM (200 mC) exhibits the highest value of pore volume of 0.0042 $cm^3 \cdot g^{-1}$, which is 10.5 times higher in comparison to that for bare nickel foam.

The wettability of the surface of the electrode is also an important parameter with respect to OER activity. A more hydrophilic surface supports the detachment of the gas bubbles that are produced during the water electrolysis process [6]. This, in turn, influences the electrocatalytical activity and stability of the catalyst in the OER process. Figure 4c presents the evolution of the contact angle for a Mn-Co film on nickel foam as a function of the deposition charge. Firstly, it can be observed that the deposition of the film for 60 mC for each studied case results in a decrease of the contact angle in comparison to that of bare nickel foam. This indicates that a thin layer of the catalytic material significantly increases the hydrophilicity of the electrode. The lowest value is obtained for the film synthesised in the solution of Mn/Co 2 mM/8 mM. A clear increase in the contact angle is observed for the higher deposition charge, which suggests that a thicker layer induces hydrophobicity of the surface. Moreover, it can be seen that a higher content of the cobalt in the Mn/Co ratio tends to be more hydrophobic with an increasing deposition charge. In addition to the changes in the film composition, the crystallographic phases present at the catalyst surface can also contribute to the evolution of the contact angle.

### 3.3. Probable Synthesis Mechanism for the Electrodeposition of Mn-Co Film on Nickel

The possible electrodeposition mechanism of the formation of the Mn-Co film on nickel in an aqueous solution of $Mn(NO_3)_2 \cdot 4H_2O$ and $Co(NO_3)_2 \cdot 6H_2O$ can be described as follows: After applying a potential of more than −1 V vs. Ag/AgCl at the nickel surface, both the reaction of the nitrate $NO_3^-$ ions (Equation (1)) coming from the Mn and Co compound, and the hydrogen evolution reaction (Equation (2)) can take place [29,34].

$$NO_3^- + H_2O + 2e^- \rightarrow NO_2^- + 2OH^- \quad (1)$$

$$NO_3^- + 7H_2O + 8e^- \rightarrow NH_4^+ + 10OH^- \quad (2)$$

Both reactions lead to an increase in hydroxide $OH^-$ ions in the solution, which then start to react with the manganese $Mn^{2+}$ and cobalt $Co^{2+}$ ions already present in the synthesis solution. The interaction of the $OH^-$ with $Mn^{2+}$ and $Co^{2+}$ leads to the formation of manganese $Mn(OH)_2$ (Equation (3)) and cobalt $Co(OH)_2$ (Equation (4)) hydroxides, respectively.

$$Mn^{2+} + 2OH^- \rightarrow Mn(OH)_2 \quad (3)$$

$$Co^{2+} + 2OH^- \rightarrow Co(OH)_2 \quad (4)$$

$$Co(OH)_2 + OH^- \rightarrow CoOOH + H_2O + e^- \quad (5)$$

In the presence of the hydroxide ions, $Co^{2+}$ oxidises further to $Co^{3+}$, forming cobalt (oxy)hydroxide CoOOH (Equation (5)). The formation of (oxy)hydroxides was also observed by others [13,61].

Thus, the final form of the as-deposited Mn-Co film on nickel may consist of a mixture of $Mn(OH)_2$ and $Co(OH)_2$, $Co_{1-x}Mn_x$-$(OH)_2$, $Ni(OH)_2$ hydroxides, and CoOOH (oxy)hydroxides, which can be both in crystalline and amorphous phases (see Figure 2a). It should be noted that other combinations of metallic oxides/hydroxides in the deposited film are also possible.

The alkaline post-treatment of the as-deposited Mn-Co film on nickel foam results in the changing of the film structure. In the presence of a strongly alkaline environment, there is a partial oxidation of $Co^{2+}$ ions into $Co^{3+}$, which leads to the formation of a higher amount of cobalt (oxy)hydroxide according to Equation (5).

Besides this, the treatment of the sample with KOH also induces the formation of $Mn^{3+}$, which was not present in the as-deposited film. Thus, the possible final form of the Mn-Co film on nickel after soaking in 1 M KOH consists of the same elements as the as-deposited film but with a higher content of $Co^{3+}$ and with the presence of $Mn^{3+}$, which indicates the formation of CoOOH and MnOOH, so the typical LDH structure of the film.

### 3.4. OER Performance of Mn-Co/Ni-Based Electrocatalysts

The electrocatalytical activity of Mn-Co films on nickel substrates in the OER was investigated electrochemically by linear sweep voltammetry (LSV) in an Ar-purged 1 M KOH electrolyte solution using a 5 $mV{\cdot}s^{-1}$ scan rate. All the current values were normalised by the geometric area of the nickel electrode. Figure 5 presents the LSV and corresponding Tafel plots investigated for the Mn-Co films on Ni foam synthesised in solutions of Mn/Co 2 mM/4 mM (a, b), 2 mM/6 mM (c, d), and 2 mM/8 mM (e, f) for different deposition charges.

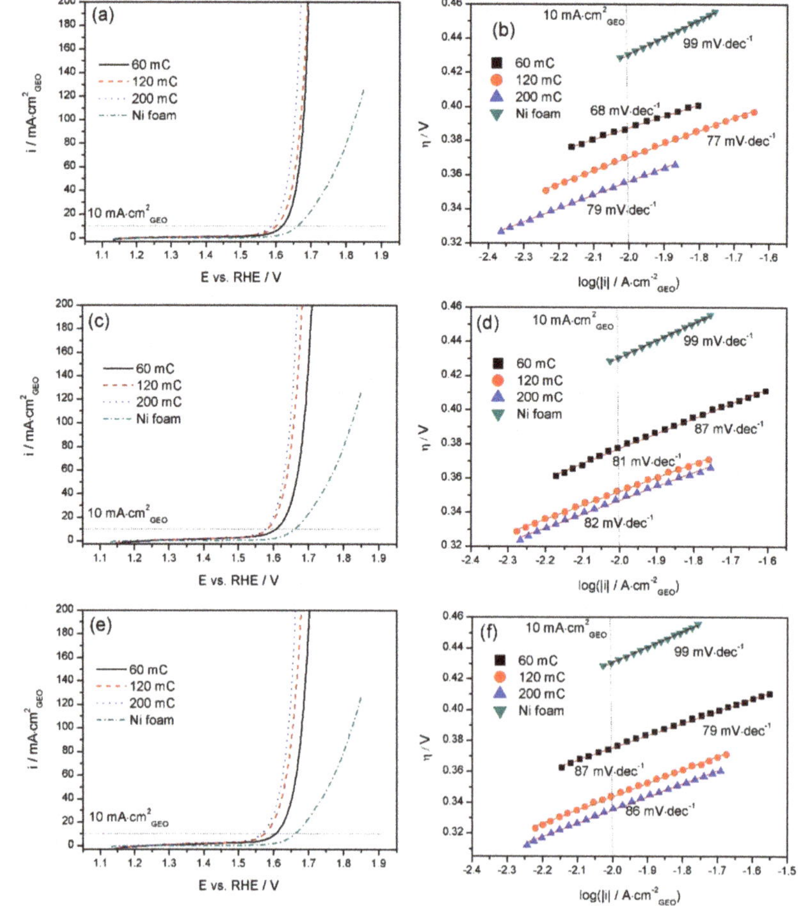


**Figure 5.** Linear sweep voltammetry profiles and corresponding Tafel plots for Mn-Co film synthesised in the presence of (**a**,**b**) Mn/Co 2 mM/4 mM, (**c**,**d**) 2 mM/6 mM, and (**e**,**f**) 2 mM/8 mM on nickel foam measured in Ar-purged 1 M KOH.

The LSV graphs show that in each studied case, nickel coated with Mn-Co revealed higher electrocatalytic properties for oxygen evolution compared to bare nickel foam, which was also assessed by the evolution of the onset potential ($E_{onset}$) and overpotential ($\eta$) of the oxygen evolution reaction (Figure 6a,b, respectively).

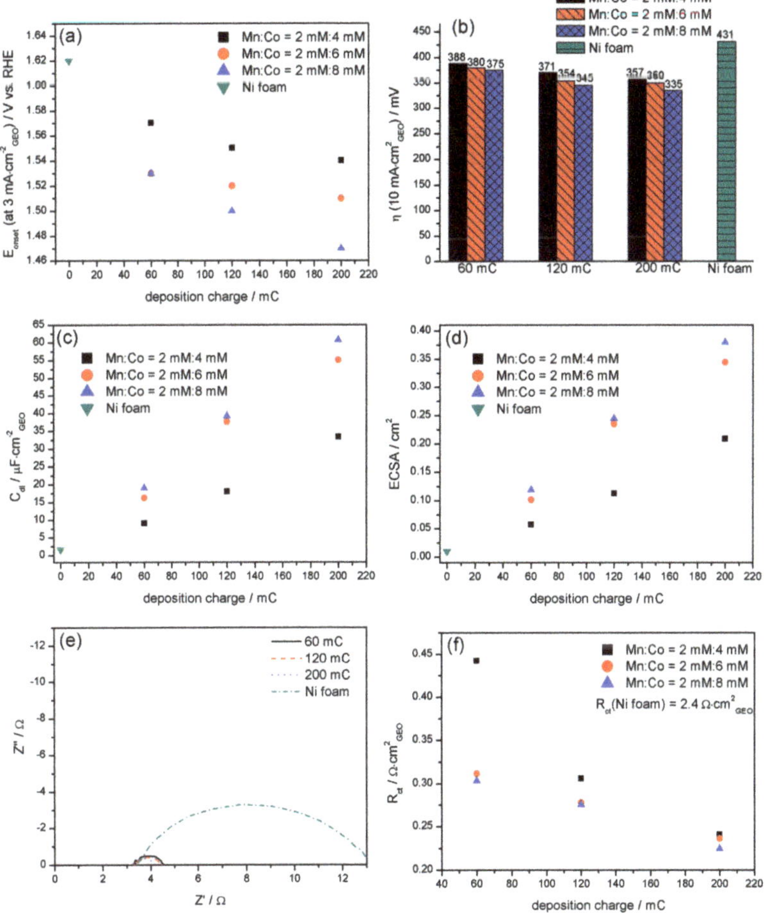


**Figure 6.** Evolution of the (**a**) onset potential, (**b**) overpotential determined at 10 mA·cm$^{-2}$, (**c**) double-layer capacitance, (**d**) electroactive surface area (ECSA), (**e**) electrochemical impedance spectra (EIS) measured at 0.7 V vs. Ag/AgCl, and (**f**) $R_{ct}$ of the oxygen evolution reaction (OER) as a function of the deposition charge for the Mn-Co film synthesised under different conditions on nickel foam.

The onset potential, which is an indication of the beginning of the oxygen evolution reaction, decreases over the deposition charge for each Mn/Co concentration ratio (Figure 6a). Moreover, the higher the content of cobalt in the Mn/Co concentration ratio in the synthesis solution, the lower the $E_{onset}$. The same trend was observed in the case of overpotential, the evolution of which is presented in Figure 6b. It should be noted that the overpotential of the OER was determined here at a current density of 10 mA·cm$^{-2}$, which is considered in the literature to be the most relevant value for solar fuel synthesis [50]. The lowest $E_{onset}$ and $\eta$(10 mA·cm$^{-2}{}_{geo}$) of 1.47 V and 335 mV vs. RHE, respectively, were obtained for the Mn-Co film synthesised in the presence of Mn/Co 2 mM/8 mM for 200 mC.

For comparison, the $E_{onset}$ and $\eta(10\ mA\cdot cm^{-2}{}_{geo})$ for bare nickel foam were determined to be 1.62 V and 431 mV vs. RHE, respectively. Moreover, the OER overpotential for commercial $IrO_2$ was determined in the literature to be 339 mV at 10 $mA\cdot cm^{-2}{}_{geo}$ in 1 M KOH [62]. The high electrocatalytic properties of the Mn-Co film synthesised under such conditions can be related to its optimised composition, which was characterised by a high specific surface area (Figure 4a) and porosity (Figure 4b) in comparison to those of the rest of the studied samples.

Because the thicker film with the higher content of cobalt versus manganese revealed the highest electrocatalytic properties for the OER, a Mn-Co film was also electrosynthesised in the presence of Mn/Co 2 mM/8 mM for 400 mC and in the presence of a higher cobalt content, i.e., Mn/Co 2 mM/12 mM for 200 mC. The LSV results showed that further changes of the deposition charge and Mn/Co concentration ratio did not also influence the catalytic properties for the OER (data not shown here).

To assess the OER mechanism, corresponding Tafel plots for bare Ni and Mn-Co films on Ni foam electrodeposited in the presence of Mn/Co 2 mM/4 mM, 2 mM/6 mM, and 2 mM/8 mM were investigated, and are presented in Figure 5b,d,f, respectively. The Tafel slopes for Ni foam were determined to be 99 $mV\cdot dec^{-1}$, which is similar to those obtained in 1 M KOH in the literature [63,64]. The slopes obtained for the Mn-Co film synthesised under different conditions reveal similar values in the range of 68–86 $mV\cdot dec^{-1}$, which suggests that the materials exhibit similar mechanisms or pathways of OER catalysis [50]. A lower Tafel slope means fast kinetics in the OER [24]. The Mn-Co film electrodeposited potentiostatically from an aqueous solution of cobalt and manganese nitrates significantly reduces such a parameter compared to bare nickel foam. Coating the metal with the Mn-Co films resulted in a modification of the nature of the active sites and thus the rate-determining step, leading to a more efficient OER [37].

To explore the intrinsic activity of the catalysts, the electrochemical active surface areas (ECSA) of Mn-Co/Ni were evaluated from the electrochemical double-layer capacitance ($C_{dl}$). Figure 6c,d present the evolution of $C_{dl}$ and the corresponding ECSA, respectively, as a function of the deposition charge for differently synthesised Mn-Co films on nickel foam. The graphs clearly show an increase in $C_{dl}$ with both the deposition charge and content of cobalt with respect to manganese in the synthesis solution (Figure 6c). The trend indicates that Mn-Co films on nickel foam prepared in the presence of a higher cobalt content and with a higher deposition charge exhibit a higher electroactive surface area. Here, it should also be noted that the value of the electrochemical surface area may not be identical to the surface area obtained from nitrogen adsorption (BET) (Figure 4a) [50]. The highest ECSA of 0.38 $cm^2$ was obtained for the Mn/Co film synthesised in the solution of Mn/Co 2 mM/8 mM for 200 mC, while the lowest ECSA of 0.05 $cm^2$ was obtained for the catalyst prepared in the presence of Mn/Co 2 mM/4 mM for 60 mC. All of the studied catalyst materials exhibited a higher surface area compared to bare nickel foam ($C_{dl}$ = 1.6 $\mu F\cdot cm^{-2}{}_{geo}$, ECSA = 0.01 $cm^2$).

In order to investigate the kinetics of the OER process, EIS measurements of the Mn-Co samples were performed in Ar-purged 1 M KOH at 0.7 V vs. Ag/AgCl. Figure 6e presents examples of the Nyquist plots for Mn-Co films synthesised in the presence of Mn/Co 2 mM/8 mM for different deposition charges. The graphs clearly show the change of the shape of the EIS spectra after coating the substrate with the catalyst. In order to study the evolution of $R_{ct}$ over the deposition charge in more detail, the EIS spectra were fitted with a simple Randles circuit with the solution resistance Rs, the charge transfer resistance $R_{ct}$, and the constant phase element CPE (Figure 6e inset). The results indicate that the lowest $R_{ct}$ of 0.22 $\Omega\cdot cm^2$ is obtained for the catalyst synthesised in the presence of Mn/Co 2 mM/8 mM for 200 mC (Figure 6f), which is ~11 times lower compared to that of bare nickel foam. A lower charge transfer resistance indicates much faster reaction rates for the OER. The acquired data are in agreement with the previous $E_{onset}$, $\eta(10\ mA\cdot cm^{-2}{}_{geo})$, BET surface area, $C_{dl}$, and ECSA analyses.

Additionally, the electrocatalytic parameters of the Mn-Co film synthesised in the presence of only cobalt nitrates ($E_{onset}$ = 1.55 V, $\eta(10\ mA\cdot cm^{-2}{}_{geo})$ = 375 mV) or in the solution with the higher content of manganese with respect to cobalt ($E_{onset}$ = 1.54 V, $\eta(10\ mA\cdot cm^{-2}{}_{geo})$ = 375 mV) for 200 mC were determined based on the LSV curves (Figure S7, Supplementary content). The results reveal their worse electrocatalytic properties for the OER compared to the Mn-Co film synthesised in a solution of Mn/Co 2 mM/4 mM, 2 mM/6 mM, or 2 mM/8 mM for 200 mC (Figure 6a,b). In spite of this, their performance is still much better in comparison to that of the bare nickel foam. This indicates that either too high a concentration content of manganese with respect to cobalt or a lack of manganese in the synthesis solution inhibits the electrocatalytic activity of the material in oxygen evolution. Only an appropriate concentration ratio of manganese to cobalt with a higher content of the latter induces high electrocatalytic properties of the material for the OER. Such findings might be due to the high electroactive surface area of the catalyst resulting from the nanosheet-like structure, which is created only in the presence of both manganese and cobalt oxides/hydroxides in a certain concentration ratio.

For comparison, LSV of the Mn-Co film electrodeposited on nickel foil was also performed (Figure S8, Supplementary content). The film deposited for Qd ≥ 120 mC was characterised by poor adhesion to the foil and progressive detachment from the substrate during the LSV measurements. The $\eta(10\ mA{\cdot}cm^{-2}{}_{geo})$ was determined to be 394 mV and 381 mV for 60 mC and 120 mC, respectively, which was higher compared to that for the Mn-Co deposited on nickel foam. The difference in the catalytic OER properties can be related to both the poor adhesion of the catalyst to the nickel foil and to the flat form of the nickel substrate. Probably, nickel in the form of a foam reveals a much more desirable structure for OER, which was also noticed by others [48].

Long-term stability is another key parameter for the practical application of an OER catalyst [6]. Figure 7 presents the chronopotentiometric graph recorded during the stability test of the Mn-Co film on nickel foam synthesised in a solution of Mn/Co 2 mM/8 mM for 200 mC measured in 1 M KOH by applying a current density of $10\ mA{\cdot}cm^{-2}{}_{geo}$.

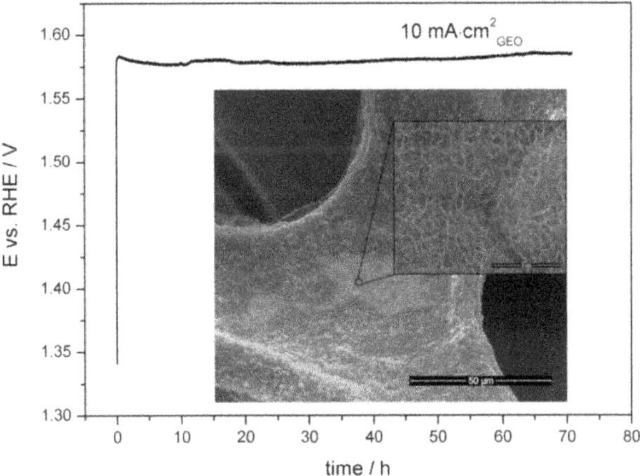


**Figure 7.** Chronopotentiometric curve recorded during the stability test of the Mn-Co film on nickel foam synthesised in a solution of Mn/Co 2 mM/8 mM for 200 mC measured in Ar-purged 1 M KOH.

The measured potential response deviated around 1.56–1.57 V vs. RHE for 70 h, which corresponds to an overpotential of 330–340 mV. The morphology of the Mn-Co film after the stability test did not change (inset of Figure 7). There was no delamination of the catalyst from the nickel substrate after removing it from the solution. All of these indicate a high electrocatalytic stability of the proposed material during the active oxygen evolution process.

The Mn-Co films electrodeposited on nickel foam synthesised in the presence of $Mn(NO_3)_2{\cdot}4H_2O$ and $Co(NO_3)_2{\cdot}6H_2O$ proposed in this work reveal promising electrocatalytic properties for the OER compared to bare nickel foam. The excellent catalytic properties can be attributed to several factors. First of all, the structure of the film consists of different kinds of mixed Mn-Co oxides/hydroxides and (oxy)hydroxides (Figure 2), which provide different possibilities of multiple redox reactions of the combined Mn-Co with hydroxide ions, resulting in enhanced catalytic activity for the OER. For comparison, the performance of the catalyst based only on cobalt oxides/hydroxides was much worse, which could be seen as both a much higher OER onset potential and overpotential (Figure S7, Supplementary content). Secondly, the morphology of the Mn-Co films was characterised by interconnected nanoflakes forming a porous interconnected 3D network (Figure 1b–d), which is desirable for ion exchange reactions. Such a structure is only formed when both Mn and Co compounds are present in the synthesis solution. This indicates that both elements significantly influence the final structure of the catalyst. The SEM results also showed that a higher amount of Co nitrates in the synthesis solution did not further influence the morphology of the deposit (Figure S4, Supplementary content). On the other hand, it was shown that both $C_{dl}$ and ECSA increased with a higher content of cobalt in the Mn/Co ratio, which in turn resulted in a lower $\eta(10\ mA\cdot cm^{-2}{}_{geo})$, $E_{onset}$, and $R_{ct}$ of the OER of the catalyst. This may indicate that the introduction of a higher amount of cobalt in the catalyst structure provides a greater number of active sites responsible for an effective catalytic reaction. Therefore, it can be suggested that Co compounds in the catalyst structure, especially the highly active crystalline β-form of $Co(OH)_2$, are mainly responsible for the enhancement of the catalytic properties of the electrode. The strong electrocatalytic properties of β-$Co(OH)_2$ were also observed by other authors [65].

The strongest electrocatalytic properties for the OER among the studied catalysts were found for the Mn-Co film electrodeposited on nickel foam in an aqueous solution of 2 mM $Mn(NO_3)_2{\cdot}4H_2O$ and 8 mM $Co(NO_3)_2{\cdot}6H_2O$ for 200 mC. The catalyst was characterised by an $E_{onset}$ of 1.47 V vs. RHE and $\eta(10\ mA\cdot cm^{-2}{}_{geo})$ of 335 mV, which indicates much higher catalytic performance for the OER in comparison to the other Mn-Co-based catalysts synthesised by chemical methods [15,62,63]. For example, the Mn-Co-based catalyst synthesised chemically was characterised by a $\eta(10\ mA\cdot cm^{-2}{}_{geo})$ of 500–600 mV in 0.1 M KOH [15]. In other work, chemically fabricated Mn-Co oxide [66] or Mn-Co oxide doped with Ce [67] exhibited $\eta(10\ mA\cdot cm^{-2}{}_{geo})$ of 450 mV and 390 mV, respectively, in 1 M KOH.

The Mn-Co-based catalyst evaluated in this work is also one of the best Mn- and/or Co-based catalysts synthesised electrochemically for the OER available in the literature (Table S1, Supplementary content). It also exhibits a lower overpotential compared to commercial $IrO_2$ (339 mV at 10 $mA\cdot cm^{-2}{}_{geo}$) determined in 1 M KOH [62]. The Mn-Co film synthesised under such conditions revealed the highest specific surface area (BET) and electrochemical surface area (ECSA) compared to the rest of the studied catalysts. The higher surface area provides a greater number of active sites, which allows for more efficient reactions with $OH^-$. The results indicate that the surface area of the catalyst has a huge impact on the OER activity. The superior stability of the Mn-Co film on nickel foam can be related not only to the type of structure and morphology but also to the lack of a polymer binder. The latter was provided by the choice of electrochemical deposition as a synthesis method, which allowed for the direct deposition of the catalyst on nickel foam providing high adhesion of the film to the substrate.

## 4. Conclusions

A Mn-Co-based film has been successfully directly deposited on nickel foam during an electrochemical deposition process in the presence of only $Mn(NO_3)_2{\cdot}4H_2O$ and $Co(NO_3)_2{\cdot}6H_2O$. A SEM analysis showed that the morphology of the Mn-Co film on nickel foam was characterised by an interconnected 3D nanoflake structure with high porosity. XRD and XPS analyses showed that the as-deposited catalysts consisted mainly of oxides/hydroxides and/or (oxy)hydroxides based on $Mn^{2+}$, $Co^{2+}$, and $Co^{3+}$. The alkaline treatment of the film in 1 M KOH resulted in the partial oxidation of the $Co^{2+}$ to $Co^{3+}$ and the creation of $Mn^{3+}$, leading to the formation of Mn-Co (oxy)hydroxides. Moreover, the XRD and TEM analyses showed the formation of crystalline $Co(OH)_2$ with a hexagonal platelet-like shape structure. These all indicate that the final form of the catalyst is based on the LDH structure, which is highly desirable for efficient OER performance.

The electrodeposited Mn-Co film on nickel foam was found to be a very promising electrocatalyst for the OER. The catalytic performance was dependent on the concentration of Mn versus Co in the synthesis solution and on the deposition charge. The optimised catalyst was obtained for Mn/Co 2 mM/8 mM for the deposition time limited by a charge of 200 mC. It was characterised by a specific surface area of 10.5 $m^2 \cdot g^{-1}$, a pore volume of 0.0042 $cm^3 \cdot g^{-1}$, and high electrochemical stability with an overpotential deviation around 330–340 mV at 10 $mA \cdot cm^{-2}{}_{geo}$ for 70 h.

**Supplementary Materials:** The following are available online at http://www.mdpi.com/1996-1944/13/11/2662/s1. Figure S1: Cyclic voltammograms recorded during sweeping the potential from 0.15 to 0.25 V vs. Ag/AgCl with different scan rates in an aqueous solution of 1 M KOH for Mn-Co film synthesised in solutions of Mn/Co 2:4 mM (a), 2:6 mM (c), and 2:8 mM (e) for 200 mC. Corresponding linear approximation of the capacitive currents versus scan rate obtained from cyclic voltammograms for Mn-Co film synthesised in solutions of Mn/Co 2:4 mM (b), 2:6 mM (d), and 2:8 mM (f) for different deposition charges. Figure S2: Synthesis graphs recorded during the potentiostatic deposition of Mn/Co oxide/hydroxides at −1.1 V vs. Ag/AgCl in aqueous solutions of differently concentrated $Mn(NO_3)_2 \cdot 4H_2O$ and $Co(NO_3)_2 \cdot 6H_2O$ with electropolymerisation time limited by a charge of 200 mC. Figure S3: Optical microscopy image of the as-deposited and after-alkaline-treatment-in-1 M-KOH Mn-Co film on nickel foam. Figure S4: SEM images of Mn-Co film synthesised in an aqueous solution of Mn/Co 2:4 mM for 60 mC (a), 2:4 mM for 120 mC (b), and 2:4 mM 200 mC (c) and 2:6 mM 200 mC (d) on nickel foam. Figure S5: SEM images of Mn-Co film synthesised in aqueous solution of Mn/Co 2:4 mM for 60 mC (a) and 2:8 mM for 200 mC (b) on nickel foil. Figure S6: SEM images of Mn-Co film synthesised in an aqueous solution of 4 mM $Co(NO_3)_2 \cdot 6H_2O$ (a) or 4 mM $Mn(NO_3)_2 \cdot 4H_2O$ and 2 mM $Co(NO_3)_2 \cdot 6H_2O$ on nickel foam for 200 mC. Figure S7: Linear sweep voltammetry profiles (a) and corresponding Tafel plots (b) of Mn-Co film synthesised in solutions of Mn/Co 2:4 mM, 4:2 mM, and 0:4 mM on nickel foam measured in Ar-purged 1 M KOH. Figure S8: Linear sweep voltammetry profiles of nickel foil and Mn-Co film synthesised in a solution of Mn/Co 2:8 mM for 60 and 120 mC on nickel foil. Table S1: Comparison of catalyst based on Mn and/or Co transition metals synthesised electrochemically for OER activity available in the literature.

**Author Contributions:** Investigation, K.C., M.K.R., G.C., J.K., and M.Ł.; validation, K.C.; writing—original draft preparation, K.C.; writing—review and editing, P.J. and S.M.; supervision, P.J. and S.M. All authors have read and agreed to the published version of the manuscript.

**Funding:** This research was funded by the Foundation for Polish Science, grant number POIR.04.04.00-00-42E9/17-00, First TEAM programme, "Nanocrystalline ceramic materials for efficient electrochemical energy conversion". K. Cysewska acknowledges the Foundation for Polish Science START stipend.

**Conflicts of Interest:** The authors declare no conflict of interest.

## References

1. Da Silva Veras, T.; Mozer, T.S.; da Costa Rubim Messeder dos Santos, D.; da Silva César, A. Hydrogen: Trends, production and characterization of the main process worldwide. *Int. J. Hydrogen Energy* **2017**, *42*, 2018–2033. [CrossRef]
2. Hosseini, S.E.; Wahid, M.A. Hydrogen production from renewable and sustainable energy resources: Promising green energy carrier for clean development. *Renew. Sustain. Energy Rev.* **2016**, *57*, 850–866. [CrossRef]
3. Bodner, M.; Hofer, A.; Hacker, V. $H_2$ generation from alkaline electrolyzer. *Wiley Interdiscip. Rev. Energy Environ.* **2015**, *4*, 365–381. [CrossRef]
4. Guo, J.; Li, Y. Ni Foam-supported Fe-Doped β-$Ni(OH)_2$ nanosheets show ultralow overpotential for oxygen evolution reaction. *ACS Energy Lett.* **2019**, *4*, 622–628. [CrossRef]
5. Mitra, D.; Trinh, P.; Malkhandi, S.; Mecklenburg, M.; Heald, S.M.; Balasubramanian, M.; Narayanan, S.R. An efficient and robust surface-modified iron electrode for oxygen evolution in alkaline water electrolysis. *J. Electrochem. Soc.* **2018**, *165*, F392–F400. [CrossRef]
6. Zeng, K.; Zhang, D. Recent progress in alkaline water electrolysis for hydrogen production and applications. *Prog. Energy Combust. Sci.* **2010**, *36*, 307–326. [CrossRef]
7. Colli, A.N.; Girault, H.H.; Battistel, A. Non-Precious electrodes for practical alkaline water electrolysis. *Materials* **2019**, *12*, 1336. [CrossRef]
8. Roger, I.; Shipman, M.A.; Symes, M.D. Earth-abundant catalysts for electrochemical and photoelectrochemical water splitting. *Nat. Rev. Chem.* **2017**, *1*, 3. [CrossRef]
9. Li, J.; Zheng, G. One-dimensional earth-abundant nanomaterials for water-splitting electrocatalysts. *Adv. Sci.* **2017**, *4*, 1600380. [CrossRef]

10. Shen, C.; Xu, H.; Liu, L.; Hu, H.; Chen, S.; Su, L.; Wang, L. Facile One-step dynamic hydrothermal synthesis of spinel $limn_2o_4$/carbon nanotubes composite as cathode material for lithium-ion batteries. *Materials* **2019**, *12*, 4123. [CrossRef]
11. Xu, J.; Zhang, H.; Xu, P.; Wang, R.; Tong, Y.; Lu, Q.; Gao, F. In situ construction of hierarchical Co/MnO@graphite carbon composites for highly supercapacitive and OER electrocatalytic performances. *Nanoscale* **2018**, *10*, 13702–13712. [CrossRef] [PubMed]
12. Tan, H.T.; Sun, W.; Wang, L.; Yan, Q. 2D Transition metal Oxides/Hydroxides for energy-storage applications. *Chem. Nano. Mat.* **2016**, *2*, 562–577. [CrossRef]
13. Nguyen, T.; Boudard, M.; Carmezim, M.J.; Montemor, M.F. Layered $Ni(OH)_2$-$Co(OH)_2$ films prepared by electrodeposition as charge storage electrodes for hybrid supercapacitors. *Sci. Rep.* **2017**, *7*, 39980. [CrossRef] [PubMed]
14. Song, F.; Bai, L.; Moysiadou, A.; Lee, S.; Hu, C.; Liardet, L.; Hu, X. Transition metal oxides as electrocatalysts for the oxygen evolution reaction in alkaline solutions: An application-inspired renaissance. *J. Am. Chem. Soc.* **2018**, *140*, 7748–7759. [CrossRef]
15. Menezes, P.W.; Indra, A.; Sahraie, N.R.; Bergmann, A.; Strasser, P.; Driess, M. Cobalt-manganese-based spinels as multifunctional materials that unify catalytic water oxidation and oxygen reduction reactions. *ChemSusChem* **2015**, *8*, 164–167. [CrossRef]
16. Li, C.; Han, X.; Cheng, F.; Hu, Y.; Chen, C.; Chen, J. Phase and composition controllable synthesis of cobalt manganese spinel nanoparticles towards efficient oxygen electrocatalysis. *Nat. Commun.* **2015**, *6*, 1–8. [CrossRef]
17. Kong, L.-B.; Lu, C.; Liu, M.-C.; Luo, Y.-C.; Kang, L.; Li, X.; Walsh, F.C. The specific capacitance of sol–gel synthesised spinel $MnCo_2O_4$ in an alkaline electrolyte. *Electrochim. Acta* **2014**, *115*, 22–27. [CrossRef]
18. Liu, M.-C.; Kong, L.-B.; Lu, C.; Li, X.-M.; Luo, Y.-C.; Kang, L. A Sol–gel process for fabrication of $NiO/NiCo_2O_4Co_3O_4$ composite with improved electrochemical behavior for electrochemical capacitors. *ACS Appl. Mater. Interfaces* **2012**, *4*, 4631–4636. [CrossRef]
19. Zhang, Y.; Xuan, H.; Xu, Y.; Guo, B.; Li, H.; Kang, L.; Han, P.; Wang, D.; Du, Y. One-step large scale combustion synthesis mesoporous $MnO_2/MnCo_2O_4$ composite as electrode material for high-performance supercapacitors. *Electrochim. Acta* **2016**, *206*, 278–290. [CrossRef]
20. Huang, T.; Zhao, C.; Wu, L.; Lang, X.; Liu, K.; Hu, Z. 3D network-like porous $MnCo_2O_4$ by the sucrose-assisted combustion method for high-performance supercapacitors. *Ceram. Int.* **2017**, *43*, 1968–1974. [CrossRef]
21. Yuan, Y.; He, G.; Zhu, J. A Facile Hydrothermal Synthesis of a $MnCo_2O_4$ reduced graphene oxide nanocomposite for applicationin supercapacitors. *Chem. Lett.* **2014**, *43*, 83–85. [CrossRef]
22. Gao, X.; Zhang, H.; Li, Q.; Yu, X.; Hong, Z.; Zhang, X.; Liang, C.; Lin, Z. Hierarchical $NiCo_2O_4$ hollow microcuboids as bifunctional electrocatalysts for overall water-splitting. *Angew. Chem.* **2016**, *128*, 6398–6402. [CrossRef]
23. Ma, W.; Ma, R.; Wang, C.; Liang, J.; Liu, X.; Zhou, K.; Sasaki, T. A superlattice of alternately stacked Ni-Fe hydroxide nanosheets and graphene for efficient splitting of water. *ACS Nano* **2015**, *9*, 1977–1984. [CrossRef] [PubMed]
24. Yan, F.; Guo, D.; Kang, J.; Liu, L.; Zhu, C.; Gao, P.; Zhang, X.; Chen, Y. Fast fabrication of ultrathin CoMn LDH nanoarray as flexible electrode for water oxidation. *Electrochim. Acta* **2018**, *283*, 755–763. [CrossRef]
25. Li, G.; Yang, D.; Chuang, P.-Y. Defining nafion ionomer roles for enhancing alkaline oxygen evolution electrocatalysis. *ACS Catal.* **2018**, *8*, 11688–11698. [CrossRef]
26. Sahoo, S.; Naik, K.K.; Rout, C.S. Electrodeposition of spinel $MnCo_2O_4$ nanosheets for supercapacitor applications. *Nanotechnology* **2015**, *26*, 455401. [CrossRef] [PubMed]
27. Jagadale, A.D.; Guan, G.; Li, X.; Du, X.; Ma, X.; Hao, X.; Abudula, A. Ultrathin nanoflakes of cobalt-manganese layered double hydroxide with high reversibility for asymmetric supercapacitor. *J. Power Sources* **2016**, *306*, 526–534. [CrossRef]
28. Wu, L.K.; Hu, J.M. A silica co-electrodeposition route to nanoporous $Co_3O_4$ film electrode for oxygen evolution reaction. *Electrochim. Acta* **2014**, *116*, 158–163. [CrossRef]
29. Nguyen, T.; Boudard, M.; Rapenne, L.; Chaix-Pluchery, O.; Carmezim, M.J.; Montemor, M.F. Structural evolution, magnetic properties and electrochemical response of $MnCo_2O_4$ nanosheet films. *RSC Adv.* **2015**, *5*, 27844–27852. [CrossRef]

30. Xiao, C.; Zhang, X.; MacFarlane, D.R. Dual-$MnCo_2O_4$/Ni electrode with three-level hierarchy for high-performance electrochemical energy storage. *Electrochim. Acta* **2018**, *280*, 55–61. [CrossRef]
31. Pan, G.T.; Chong, S.; Yang, T.C.K.; Huang, C.M. Electrodeposited porous $Mn_{1.5}Co_{1.5}O_4$/Ni composite electrodes for high-voltage asymmetric supercapacitors. *Materials* **2017**, *10*, 370. [CrossRef] [PubMed]
32. Clark, M.; Ivey, D.G. Nucleation and growth of electrodeposited Mn oxide rods for supercapacitor electrodes. *Nanotechnology* **2015**, *26*, 384001. [CrossRef]
33. Vigil, J.A.; Lambert, T.N.; Eldred, K. Electrodeposited $MnO_x$/PEDOT composite thin films for the oxygen reduction reaction. *ACS Appl. Mater. Interfaces* **2015**, *7*, 22745–22750. [CrossRef] [PubMed]
34. Maile, N.C.; Fulari, V. Electrochemical synthesis of $Mn(OH)_2$ and $Co(OH)_2$ thin films for supercapacitor electrode application. *Aarhat Multidiscip. Int. Educ. Res. J.* **2018**, *48178*, 48818.
35. Ramírez, A.; Hillebrand, P.; Stellmach, D.; May, M.M.; Bogdanoff, P.; Fiechter, S. Evaluation of $MnO_x$, $Mn_2O_3$, and $Mn_3O_4$ electrodeposited films for the oxygen evolution reaction of water. *J. Phys. Chem. C* **2014**, *118*, 14073–14081. [CrossRef]
36. Castro, E.B.; Gervasi, C.A.; Vilche, J.R. Oxygen evolution on electrodeposited cobalt oxides. *J. Appl. Electrochem.* **1998**, *28*, 835–841. [CrossRef]
37. Pérez-Alonso, F.J.; Adán, C.; Rojas, S.; Peña, M.A.; Fierro, J.L.G. Ni/Fe electrodes prepared by electrodeposition method over different substrates for oxygen evolution reaction in alkaline medium. *Int. J. Hydrogen Energy* **2014**, *39*, 5204–5212. [CrossRef]
38. Han, S.; Liu, S.; Wang, R.; Liu, X.; Bai, L.; He, Z. One-Step Electrodeposition of Nanocrystalline $Zn_xCo_{3-x}O_8$ Films with High Activity and Stability for Electrocatalytic Oxygen Evolution. *ACS Appl. Mater. Interfaces* **2017**, *9*, 17186–17194. [CrossRef]
39. Lambert, T.N.; Vigil, J.A.; White, S.E.; Davis, D.J.; Limmer, S.J.; Burton, P.D.; Coker, E.N.; Beechem, T.E.; Brumbach, M.T. Electrodeposited $Ni_xCo_3$ nanostructured films as bifunctional oxygen electrocatalysts. *Chem. Commun.* **2015**, *51*, 9511–9514. [CrossRef]
40. Koza, J.A.; He, Z.; Miller, A.S.; Switzer, J.A. Electrodeposition of crystalline $Co_3O_4$-A catalyst for the oxygen evolution reaction. *Chem. Mater.* **2012**, *24*, 3567–3573. [CrossRef]
41. Liu, P.F.; Yang, S.; Zheng, L.R.; Zhang, B.; Yang, H.G. Electrochemical etching of α-cobalt hydroxide for improvement of oxygen evolution reaction. *J. Mater. Chem. A* **2016**, *4*, 9578–9584. [CrossRef]
42. Castro, E.B.; Gervasi, C.A. Electrodeposited Ni-Co-oxide electrodes: Characterization and kinetics of the oxygen evolution reaction. *Int. J. Hydrogen Energy* **2000**, *25*, 1163–1170. [CrossRef]
43. Kim, T.W.; Woo, M.A.; Regis, M.; Choi, K.S. Electrochemical synthesis of spinel type $ZnCo_2O_4$ electrodes for use as oxygen evolution reaction catalysts. *J. Phys. Chem. Lett.* **2014**, *5*, 2370–2374. [CrossRef]
44. Bao, J.; Xie, J.; Lei, F.; Wang, Z.; Liu, W.; Xu, L.; Guan, M.; Zhao, Y.; Li, H. Two-dimensional Mn-Co LDH/Graphene composite towards high-performance water splitting. *Catalysts* **2018**, *8*, 350. [CrossRef]
45. Jia, G.; Hu, Y.; Qian, Q.; Yao, Y.; Zhang, S.; Li, Z.; Zou, Z. Formation of Hierarchical structure composed of (Co/Ni)Mn-LDH nanosheets on MWCNT backbones for efficient electrocatalytic water oxidation. *ACS Appl. Mater. Interfaces* **2016**, *8*, 14527–14534. [CrossRef] [PubMed]
46. Lankauf, K.; Cysewska, K.; Karczewski, J.; Mielewczyk-Gryn, A.; Górnicka, K.; Cempura, G.; Chen, M.; Jasiński, P.; Molin, S. $Mn_xCo_{3-x}O_4$ spinel oxides as efficient oxygen evolution reaction catalysts in alkaline media. *Int. J. Hydrogen Energy* **2020**, *45*, 14867–14879. [CrossRef]
47. Qiao, Y.; Jiang, K.; Deng, H.; Zhou, H. A high-energy-density and long-life lithium-ion battery via reversible oxide–peroxide conversion. *Nat. Catal.* **2019**, *2*, 1035–1044. [CrossRef]
48. Grdeń, M.; Alsabet, M.; Jerkiewicz, G. Surface science and electrochemical analysis of nickel foams. *ACS Appl. Mater. Interfaces* **2012**, *4*, 3012–3021. [CrossRef]
49. Jung, S.; McCrory, C.C.L.; Ferrer, I.M.; Peters, J.C.; Jaramillo, T.F. Benchmarking nanoparticulate metal oxide electrocatalysts for the alkaline water oxidation reaction. *J. Mater. Chem. A* **2016**, *4*, 3068–3076. [CrossRef]
50. Yu, J.; Zhong, Y.; Zhou, W.; Shao, Z. Facile synthesis of nitrogen-doped carbon nanotubes encapsulating nickel cobalt alloys 3D networks for oxygen evolution reaction in an alkaline solution. *J. Power Sources* **2017**, *338*, 26–33. [CrossRef]
51. Olivier, J.P.; Conklin, W.B.; Szombathely, M.V. Determination of pore size distribution from density functional theory: A comparison of nitrogen and argon results. *Stud. Surf. Sci. Catal.* **1994**, *87*, 81–89. [CrossRef]

52. Liu, S.; Lee, S.C.; Patil, U.; Shackery, I.; Kang, S.; Zhang, K.; Park, J.H.; Chung, K.Y.; Chan Jun, S. Hierarchical MnCo-layered double hydroxides@Ni$(OH)_2$ core–shell heterostructures as advanced electrodes for supercapacitors. *J. Mater. Chem. A* **2017**, *5*, 1043–1049. [CrossRef]
53. Li, R.; Hu, Z.; Shao, X.; Cheng, P.; Li, S.; Yu, W.; Lin, W.; Yuan, D. Large scale synthesis of NiCo layered double hydroxides for superior asymmetric electrochemical capacitor. *Sci. Rep.* **2016**, *6*, 18737. [CrossRef]
54. Liu, Z.; Ma, R.; Osada, M.; Takada, K.; Sasaki, T. Selective and controlled synthesis of α- and β-cobalt hydroxides in highly developed hexagonal platelets. *J. Am. Chem. Soc.* **2005**, *127*, 13869–13874. [CrossRef] [PubMed]
55. Zhao, X.; Niu, C.; Zhang, L.; Guo, H.; Wen, X.; Liang, C.; Zeng, G. Co-Mn layered double hydroxide as an effective heterogeneous catalyst for degradation of organic dyes by activation of peroxymonosulfate. *Chemosphere* **2018**, *204*, 11–21. [CrossRef] [PubMed]
56. Biesinger, M.C.; Payne, B.P.; Grosvenor, A.P.; Lau, L.W.M.; Gerson, A.R.; Smart, R.S.C. Resolving surface chemical states in XPS analysis of first row transition metals, oxides and hydroxides: Cr, Mn, Fe, Co and Ni. *Appl. Surf. Sci.* **2011**, *257*, 2717–2730. [CrossRef]
57. Fujiwara, M.; Matsushita, T.; Ikeda, S. Evaluation of Mn3s X-ray photoelectron spectroscopy for characterization of manganese complexes. *J. Electron Spectrosc. Relat. Phenom.* **1995**. [CrossRef]
58. Luo, X.; Lee, W.-T.; Xing, G.; Bao, N.; Yonis, A.; Chu, D.; Lee, J.; Ding, J.; Li, S.; Yi, J. Ferromagnetic ordering in Mn-doped ZnO nanoparticles. *Nanoscale Res. Lett.* **2014**, *9*, 625. [CrossRef]
59. Wu, N.; Low, J.; Liu, T.; Yu, J.; Cao, S. Hierarchical hollow cages of Mn-Co layered double hydroxide as supercapacitor electrode materials. *Appl. Surf. Sci.* **2017**, *413*, 35–40. [CrossRef]
60. Cysewska, K.; Karczewski, J.; Jasiński, P. Influence of electropolymerization conditions on the morphological and electrical properties of PEDOT film. *Electrochim. Acta* **2015**, *176*, 156–161. [CrossRef]
61. Yan, F.; Zhu, C.; Li, C.; Zhang, S.; Zhang, X.; Chen, Y. Highly stable three-dimensional nickel–iron oxyhydroxide catalysts for oxygen evolution reaction at high current densities. *Electrochim. Acta* **2017**, *245*, 770–779. [CrossRef]
62. Zhou, H.; Yu, F.; Zhu, Q.; Sun, J.; Qin, F.; Yu, L.; Ren, Z. Water splitting by electrolysis at high current densities under 1.6 volts. *Energy Environ. Sci.* **2018**, *11*, 2858–2864. [CrossRef]
63. Yang, Z.; Zhang, J.-Y.; Liu, Z.; Li, Z.; Lv, L.; Ao, X.; Tian, Y.; Zhang, Y.; Jiang, J.; Wang, C. Cuju-structured iron diselenide-derived oxide: A highly efficient electrocatalyst for water oxidation. *ACS Appl. Mater. Interfaces* **2017**, *9*, 40351–40359. [CrossRef] [PubMed]
64. Zhang, J.; Wang, T.; Liu, P.; Liao, Z.; Liu, S.; Zhuang, X.; Chen, M.; Zschech, E.; Feng, X. Efficient hydrogen production on $MoNi_4$ electrocatalysts with fast water dissociation kinetics. *Nat. Commun.* **2017**, *8*, 15437. [CrossRef]
65. Mefford, J.T.; Akbashev, A.R.; Zhang, L.; Chueh, W.C. Electrochemical reactivity of faceted β-$Co(OH)_2$ single crystal platelet particles in alkaline electrolytes. *J. Phys. Chem. C* **2019**, *123*, 18783–18794. [CrossRef]
66. Park, K.R.; Jeon, J.E.; Ali, G.; Ko, Y.-H.; Lee, J.; Han, H.; Mhin, S. Oxygen evolution reaction of Co-Mn-O electrocatalyst prepared by solution combustion synthesis. *Catalysts* **2019**, *9*, 564. [CrossRef]
67. Huang, X.; Zheng, H.; Lu, G.; Wang, P.; Xing, L.; Wang, J.; Wang, G. Enhanced water splitting electrocatalysis over mnco2o4 via introduction of suitable ce content. *ACS Sustain. Chem. Eng.* **2019**, *7*, 1169–1177. [CrossRef]

MDPI
St. Alban-Anlage 66
4052 Basel
Switzerland
Tel. +41 61 683 77 34
Fax +41 61 302 89 18
www.mdpi.com

*Materials* Editorial Office
E-mail: materials@mdpi.com
www.mdpi.com/journal/materials

www.ingramcontent.com/pod-product-compliance
Lightning Source LLC
LaVergne TN
LVHW070623170726
843515LV00004B/164
*9783036532837*